Retiming The Earth

Rethink Millions into Thousands

By Steve Preston

2nd Edition

Table of Contents

RETIMING THE EARTH 1
TABLE OF CONTENTS 3
INTRODUCTION 5
CREATIONIST THEORY PROBLEMS 13
MANY DATING PROBLEMS? 20
RADIO-ACTIVE DECAY TIMING 27
RADIOACTIVE DECAY VARIATIONS 30
ICE CORE DATING 32
PALEO-MAGNETICS AGAIN 34
HOT SPOT DATING..39
UNSTABLE ORBITS 46
MARS MESS 48
WEIRD MARTIAN CRATERS 51
PEOPLE WALKING WITH DINOSAURS 61
ANCIENT WORLD OF GIANTS....................................63
TITANIC CIVILIZATION 67
HOW BAD WAS THE EXTINCTION? 79
ODDLY SOME THINGS SURVIVED 80
ANAK AND ANAKIM 86
RESULTS OF THE HEAVEN WAR 90
CRETACEOUS EXTINCTION 98
FOSSILIZATION DATING 101
DINOSAURS WERE NOT FOSSILIZED 104
DINOSAURS REMADE 107
"BOOK OF SECRETS" DINOSAUR MAKING..............112
"THE BOOK OF GIANTS" CREATIONS 115
WAR AND WEAPONS..119
INDIAN FLYING 124
AFRICAN FLYING...125
FLYING IN THE AMERICAS..128
OTHER FLYING 130
SPACE WAR 132
VENUS 12,000 YEARS AGO 135
VENUS GOES CLOSER TO THE SUN..........................147
DINOSAURS DIE BEFORE EARTH SHIFT.................150
MANY ISLANDS FLOODED 152

FLOOD BEFORE THE FLOOD 159
WATER TEMPERATURE...163
EARTH SHIFT EVIDENCE 164
NOAH'S FLOOD MARKER 166
ANAK SURVIVED THE FLOOD 168
DINOSAURS DIED IN THE FLOOD 170
PHYSICAL EVIDENCE FLOOD TIMING.....................173
MYSTERY OF MEMORY LOSS 178
APE THEORY 192
ELEPHANT THEORY 196
LIFE SPAN CHANGE...199
BEHEMOTH...206
TRUTH...218
CONCLUSIONS...219
ABOUT THE AUTHOR...222

Introduction

I'm sorry! I have been an accomplice in trying to get people to believe that the Earth has been here and animals have been here for a longer time than the evidence proves and time between significant events before the 50 thousand year stabilization must be compressed.

This Has to Stop!

If you remember form my earlier books, I identified the time of the great Heaven War and the extinction of the dinosaurs as the same event. At the same time, I indicated that the time lag between Adam and Cro-Magnon man and this extinction event was so very long that it seemed impractical, but the evidence showed these two events being markers we could reasonably use. Recent discoveries are blasting away at theories that identify the Age of Dinosaurs from millions of years ago to tens of thousands so that the time lines can and must, now, be compressed to something that is reasonable. While I'm not suggesting the beginnings of the universe and Earth have significantly changed in timeline before the more recent events of the emergence of animal life, I do believe that we always need to consider that time itself is not a constant and descriptions of same must always be held with looseness. I also am not saying that evolution does not exist, but from my earlier books the genetic manipulation of God's creations have turned most animals into abominations of his first animals and the reason most are described as "UNCLEAN" is not that he hated his own creations. It was that the original animals had been made unclean by manipulation by scientists. The following portion of a earth history timeline shows the general nature of what I am talking about. The events prior to the 250 thousand year extinction and destruction of all continents except for Pangea, I am not considering changing for a simple reason, there doesn't seem to have been people here before that time. There are too many time riders that show the slow healing of the earth has been going on since that time, so let's just say those dates are ACCEPTIBLE. What I am considering is that after that fateful event we got things screwed up. We considered everything still as a constant and the Earth was in no way stable or similar to what it was before the Pacific Ocean was created. Now that we know many of the ancient dinosaurs lived well into the time of modern eras, we must reconsider a stable timeline. For one thing, the huge expansion of life after the 250 thousand year old event makes no sense without a lot of help. We are going to explore the help and a much more compressed timeline that goes along better with the other details of history. Each bullet represents years before the present.

450 Thousand -The Permian Age. Animals leave the water and land animals reign supreme. It would end with the large mass extinction of animals.

350 Thousand-Mars and Earth come close a first time and form the long chain of mountains that include the Himalaya Mountains. This Ends the Jurassic Period. Most Jurassic animals die.

250 Thousand -Mars and Earth come near again and mountains are formed. Pacific Ocean is sucked out of the Earth and the only remaining continent, Pangea, splits and moves to repair the hole. This ends the Triassic period- most Triassic animals die.

145,000-Ancient Humans are created and colonize several planets. Greeks called them Titans and Genesis said they were the *"Giants of Old"*. This would be the beginning of the Cretaceous Period.

135,000-Genetic manipulation by ancient humans and a challenge to make the largest and craziest looking creatures

130,000 -Angels come into existence and ancient humans begin to lose their physical bodies.

125,000-1st Heaven War is so very horrible that the Earth rotation is slowed down. This action kills the huge dinosaurs and anyone left on the earth.

100,000-Following the devastating Heaven War. A new human comes into existence called the ANAK. The Ancient humans all but vanished.

90,000-A huge change in the Earth again showed itself. A change in the decay of nuclear materials was noted as the nuclear events of the war were finally halted

80,000-ANAK try to regenerate humans but none of their creations seem to be useful to them. Greatest design was Homo-Habilis

70,000 -6th day man is created as a worker for the ANAK. [Homo-Erectus]

55,000-ANAK genetically breed "Hybrid humans" and some become giants. The planets are possibly repopulated. [Neanderthal]

40,000-Still another human is created by God along with new animals. The new type of human is called *"Adamic"* after the first human.

30,000-This new human has sex with Nephilim and hybrid children are the result.

20,000-Angels turn into another type human creature called Nephadim, have sex with humans resulting in giant children.

18 to 16K-Massive worldwide wars are fought.

12,000-The Venusian moon shatters and meteorites destroy much of the earth. Several sites similar to Atlantis are destroyed and the Venusian Dark Age begins.

8000BC-The worldwide flood comes from a comet strike and an earth axis shift happening about the same time. A few Adamics, Hybrids, and Nephilim survive.

3500BC-The Tower of Babel is built and massive worldwide wars. Something happens and humans are changed. Afterwards they can use only a small part of their brain.

3150BC- Egyptians called it ZEP-TEPI, Indians called it the Beginning of the Age of Kali, and the PreMaya started their well-known calendar.

0-God Incarnate visits

2026- Prophecies of the Moslem Wars and the ends of times to come.

This timeline is substantially compressed over what you normally see. Mainstream, well meaning, lazy, researchers pull old details of the development of the earth and mankind and provide a timeline similar to the one shown next.

Standard Geological Timeline

Era/Period/Epoch	Time (M Years ago)
Archaeozoic Period	5000-1500
Proterozoic Period	1500-545
Cambrian period	550-500
Ordovician period	500-440
Silurian period	440-410
Devonian period	410-365
Carboniferous period	365-300
Permian period	300-250
Triassic period	250-212
Jurassic period	212-145
Cretaceous period	145-65
Tertiary period	65-0.04
Pleistocene period	0.04-0.01
Holocene period	0.01-0

Wow! How neat. Hundreds of millions of years, and many, many long time periods show up to embellish our history and allow for a development timeline that allows for something we call evolution. I'm not saying evolution is bunk. There is no question that evolution is real and it caused many of the changes in characteristics of animals seen in ancient times and today. That being said, survival of the fittest, is not the reason for most of the adaptations and stretching out the timeline to allow for the level of absurdity is not called for.

The main reason scientists have allowed themselves to go down the rabbit hole of the previous timeline is something called radioactive decay. Thought to be a constant and well documented, believable, unshaking truth, we now find out that many things change these constants tremendously.

Possible Geological Timeline Modification

A better timeline might be the one shown below. Instead of hundreds and hundreds of millions of years, the timeline looks like it must be compressed to be a hundredth of that originally described as unshakable. I know this sounds absurd to you right now, but, hopefully when you read about the proof, you will be less inclined to simply take for granted that you are being told the whole truth and investigate on your own. I am not saying my timeline is unshakable either, but it explains a lot to the anomalies typically being brushed over by those trying to control reality by decree.

Era/Period/Epoch	time (T years ago)
Archaeozoic period	50,000-3000
Proterozoic period	3000-1000
Cambrian period	1000-900
Ordovician period	900-800
Silurian period	800-700
Devonian period	700-600
Carboniferous period	600-500
Permian period [1st Mars event]	500-400
Triassic period [Pacific Ocean]	400-300
Jurassic period [Titan]	300-200
Cretaceous period [ANAK]	200-100
Tertiary period [Adam]	100-40
Quaternary period [Flood]	40-10
Holocene period [Present]	10-0

Less Known Extinctions

In order to emphasize the nature of the graph, the following chart was generated from basic extinction numbers determined at each of the major extinction layers. At each extinction period, I have provided the animal diversity just before the extinction as compared with the animal diversity of today. These numbers are those established by the "normal" timeline, but the percentages hold true for this characterization as well. From it [Animal Diversity] we can determine that 500 thousand years ago there were about 50 percent as many different species as we believe are in existence today and that diversity level generally increases between then and now along a slope that almost ignores the extinction periods. This is one of the main reasons scientist conclude that the timeline MUST be lengthened, even if it makes no sense. An extremely fast rise in animal life diversity after most of the extinction periods is bewildering to those who adopt a notion that standard evolution without direction is the only plausible explanation for animal diversity. Others claim that God simply changes the millions of type of animals willy-nilly as a master controller of animal diversity, but they are stopped by most of the ancient texts that explain that MOST of the animals are UNCLEAN abominations. Certainly, the fast rise was not possible if evolution caused the increase, because there just weren't enough animals that survived the extinctions to establish critical colonies. Certainly, God did not make millions of animal types over and over again only to identify them as abominations. Something else has been going on. Humans, living during these time periods made most of the animals from God's original ones. Others evolved from God's original seed, but the idea that globs of sugar on a rock filled with water were electrified by lightning to build DNA and "SOMEHOW" this DNA became living DNA for no particular reason is highly unlikely. By the way--- here is a question---- what is the difference between living and dead DNA? What I can tell you is that it is not electricity, magnetism, gravity, accelerated time, black hole jizz or anything known to man. The whole concept of creation and genetic manipulation is not a main theme of this book, so let me halt this line of thought and get down to business.

Years ago	Boundary Name	Animal Loss
900K	Cambrian	70%
800K	Ordovician	80%
700K	Silurian	90%
600K	Devonian	75%
500K	Carboniferous	80%
400K	Permian [First Uplift]	80%
300K	Triassic [Pacific]	95%
200K	Jurassic [Heaven War]	90%
100K	Cretaceous [ANAK rule]	70%
40K	Paleocene [Adam]	85%
25K	Oligocene [Civil War]	40%
12K*	Miocene [World War]	30%
10K	Pleistocene [Flood]	80%

* Biblical and other ancient Jewish works indicate that 1/3 of all life-forms died during this time.

While I am apologizing for pushing the "more common timeframes", I will be discussing just why the timeline had to be corrected as I have done. As I just mentioned, this does not mean that evolution is not a very important element of existence. It is clear that God initially used this progression of species, but mankind wanted to control things and that started a big mess. Most evolutionists and most geologic scientists held there hat on the solid accounting of nuclear decay in timing everything from the age of the Earth to how long muck was in a back yard. Nuclear Decay changes all the time making me have to apologize for previous timeline analysis and building up a new one that agrees with all the new evidence that is now available. Let me first start with issues.

Creationist Theory Problems

In know you have at least been aware that an antagonistic group of the "Evolutionaries" is a group that addresses and uses many of the discoveries associated with time line misunderstanding. You would think they would now have a reasonable timeline to substitute in its place, but they are burdened with mistranslation of the first book of the Bible. When it talks about "Ages" like "the day of" or "harvest day" etc., they build a timeline that suggests, the entire world was "CREATED" in only six 24 hour periods. Before I get into this one, let me first say that I strongly believe that the Bible is written in an accurate and sequentially significant way, but its time base is event driven or relation driven rather than absolute time. Some try to force fit the absolute time into secular dogma and when there are problems, very interesting devices are used to reestablish the "absolute timeline". Let's look at problems with absolute timing without regard for other evidence.

The Creationist Theory has a number of problems that are pushed under the rug. The theory essentially is that the world can only be about 6 thousand years old, and the creation of all animals occurred within a 6-day period in accordance with the first chapter of the Biblical book known as Genesis. In order to keep this belief one must use an extremely limited interpretation of the Biblical texts. We will see that the evidence does not support this type of theory; nor does the Bible for that matter. Outside a small portion of the Christian Community, this theory is typically scoffed at, but aside from the extremely shortened time-period, many of the elements of its basis go along with much of the evidence, and so they will not be ignored in this history. Here are some of the elements that show the time base of this theory is, most likely, too short.

Magnetic Field Changes

Many Magnetic field changes have been verified by lava samples from the Atlantic Ocean and other sites. As the planet plates separate, lava pours out in abundance in the middle of the ocean, and the metallic portions of the lava align to the current magnetic field of the Earth before it hardens. Core samples from the bottom of the Atlantic show that there have been at least 170 major changes in the magnetic field alignment and none have occurred in the last 2 thousand years. A process was developed which uses these metal alignments for dating purposes. It is called Paleo-magnetic dating and it does not confirm a short time span. Each time that the magnetic field changes havoc and destruction is rendered on the inhabitants of the Earth. By creationist standards, the 170 shifts would have occurred in 4 thousand years of improbable and unbelievable massive wobbles, destructions, and annihilations and, during all this mess, the Biblical story has almost no mention of the upheavals except for one that occurred 7 days before the worldwide flood. We will discuss in detail the destruction periods that are revealed in the Biblical texts, but they do not account for what is found in the Atlantic.

Worldwide Flood

The worldwide flood could not have created the thick pockets of coal, as implied by the creationists. Even if the flood produced wide and uncontrolled manufacture of coal deposits, the amount and depth of the deposits around the world could not possibly have been generated during that one major Earth trauma. In some areas, the coal deposits are well over a mile thick and are over an extremely wide area. The thought that the trees all congregated in one spot as they floated around in the flood and then collected to form these massive coal deposits is not probable. Trees must have been in an area, then died, grew, died again, grew, etc. ---for many, many thousands of years. By the way, there was a worldwide flood, but it most probably happened 10 thousand years ago and this has been verified by a substantial amount of evidence. We will also discuss another, almost as devastating, flood that happened about 3 thousand years before the big one. You may have heard of this one as the sinking of Atlantis, but this is not a book to address this flood except to show how and why it did occur and the timeline of its happening.

Physical Similarities

The physical characteristics between the shorelines of South America and Africa, along with many other indications, strongly suggests that there was once one major continental mass that has been separating slowly over many, many years. It could not have happened over the 2000 years of creation before the flood without showing a major difference in the crustal density in the middle of the Atlantic Ocean. The crustal density is identical in the Atlantic and on the continental masses. Only under the Pacific Ocean is there any major variation in density of the crust, which brings us to another problem. Some suggest the Atlantic is getting larger at the same rate that is was getting larger millions of years ago. Because the Ocean is increasing in size about 1 foot every decade, the Pacific must be 200 million years old. I was one of the believers in this supposed simple and accurate timing, but it does not seem to be nearly as accurate as many had originally believed.

Thin Crust

The Earth's crust is thin under the Pacific Ocean. By using seismic mapping techniques it has been determined that the thickness of the crust at the bottom of the Pacific Ocean is much, much less thick in comparison with the thickness of the crust around the rest of the world. This strongly suggests that the Earth was split apart in ancient times and is slowly healing itself. By examining the amount of crustal matter that is deposited each year, the age during which the Earth was split apart has been estimated to be over 200 million years ago. Again we fall into the trap that suggests that deposited materials are fairly constant over time. There is no doubt that it has taken a long time to build up the bottom of the Pacific, but it is not 200 million years.

Carbon 14

Without a doubt we have found items that tested to be older than they actually were with this carbon 14 test method. After a volcanic action the amount of carbon 14 was decreased more dramatically than normal, which shifted the testing results but we have also found items that were dated to be younger than they actually were. This was noticed whenever the amount of Carbon 14 was increased unnaturally. Carbon 14 has a half-life of 5600 years. That type of dating is only good for about 40 thousand years and assumes that no outside force increase the percentage of carbon 14 on the tested sample. Not only has the 40 thousand year limit been extended greatly over the past few years, but also many variants of radiocarbon dating and other techniques are now being used to verify and cross-compare dates. While carbon 14 dating is a good comparative dating method, I have now reconciled that all Nuclear deterioration timing methods have the opportunity of great error depending on the position of the sun.

Oxygenated Air

It has been suggested that the air was more oxygenated long ago and therefore, the decay process was modified before the flood, more oxygen should mean more carbon-based items. This would mean that there would be a higher concentration of carbon 14 than we currently see, which would in-turn mean that any carbon-14 dating that crossed the high oxygen boundary would be in error---This error would indicate that things were actually much older would test to be less old because there would be too much carbon-14 remaining. Items tested to be 6,000 years old may be twice as old, which of course is the wrong direction for the creationist belief.

Nuclear Events

An additional problem for carbon-14 dating is a nuclear event, and the Earth has seen many. One such Nuclear destruction period happened between 5500 and 6 thousand years ago. It was so horrible, the Egyptian restarted their clocks 3150BC, The PreMaya did the same and started their well-known new calendar 3150 BC, To this we add the people of India who started the New Age "age of Kali" 3150BC or so. In the city of Mohen jo Daro [city of the Dead in Pakistan] hundreds of skeletal remains lined the streets. Of note, the skeletons are still radio-active after thousands of years.

In addition to the bodies, hundreds of black lumps of melted clay pots littered the streets and skeletons were found in the street holding hands as if in complete terror during the last minutes of life. Some researchers have indicated that these skeletons were the most radioactive ever found. As you might expect, stones on walls were fused together as if a nuclear explosion occurred. Carbon dating of some of the remaining skeletons indicated that they were certainly over

2500 years old, and the site itself has been determined to be much older. This topic will be readdressed later as we go through this whole new timeline.

Carbon dating after a nuclear blast is hampered due to the increased carbon isotopes that are formed. It is strongly believed that the incident above actually occurred about 5 thousand years ago during the last great World War. The reason I'm bringing this up is that the radioactive nature of the destruction "ADDED to the normal amount of radio-active C14 used in dating which made the skeletons seem younger than they really were.

I KNOW YOU ARE THINKING---"If nuclear events make things seem "LESS OLD" the "normal nuclear dating would be in error the other way and the dates from this type of dating should be reevaluated as even more ancient. What we now know is that even things like Solar flares greatly affect nuclear deterioration times the opposite way and make things appear to be MUCH OLDER. I'm getting ahead of myself so just know that nuclear dates are not constant right now and let's get back to creationist issues.

Too Many Animals

Another issue is that there have been so many animals on the Earth that we still haven't run out of the oil that was produced by the decay of their bodies. If the animals were only here during the 2 thousand years before the flood, the Earth must have somehow been much, much, much larger to allow them to walk and not be piled on top of one another. I know that some believe that oil came from some other means, but right now let's just assume that it came from dinosaurs, because it is the most logical without additional insight.

Karoo

On the southern portion of Africa lies the greatest find of terrestrial vertebrate fossils [mostly swamp dwelling reptiles].

It is estimated that there are 800 Billion; yes that's billion with a B, animals in a sandstone and shale deposit that is 20,000 feet thick.

It is stretched out for hundreds of miles. This simply could not have been a single clump of creatures pushed into one area as the floodwaters subsided 10000 years ago or only 4000 years ago as many Creationists believe.

Differences Versus Time

There are too many differences and not enough time. According to Creationist view, the flood occurred 4000 years ago and only olive skinned Adamic people survived. Within a period of about 100 years, they mutated into red skin people, white skin, yellow skin, black skin, brown skin, straight hair, curly hair, flat nose, high cheek bone, and slanted eyed variants around the world. Then, for the next 3,900 years, nothing happened at all. Carvings from thousands of years ago and today show people look the same. Those first 100 years must have been something if we are to believe a flood date of 4 thousand years ago and only Noah descendants as the survivors.

So you still may be thinking, "It sounds like the "Normal" dating of things must be at least pretty close to being right." I think for many things the dating of things is a good approximation, but let's look a little closer.

Many Dating Problems?

For this section, I'm going over some of the more well-known dating methods to see how they fare. If they can be calibrated, these should give us the same dates obtained from other methods.

Stratographic Positioning

This is the determination of age by position, depth, and material consistency. Sometimes this is the only method for cross comparison that is reasonable. Scientists simply determine the depth of objects, or features near the object, or number of lava flows, or similar geologic characteristics and use the depth as a time gage. This type of comparison may not have a very high level of accuracy, but seeing things in different layers seem to show when something died. If something is lower, it is older and newer is newer. Added to this method is something called the K-T boundary, where iridium chalk was deposited from an ancient meteor that struck the Yucatan around the time the dinosaurs died. Scientists have been using this for a long time when, all of a sudden, there were trees found that were going the wrong way. The next set of pictures shows some of the unfortunate trees that must have died repeatedly to be deposited perpendicular to all of the stratographic lines.

These things have been found all over including Germany, France, UK, California, several eastern States and Nova Scotia so they had to even name this oddness as Polystrate fossils. Because they are standing, this means that the layers of sediment formed around the thing over whatever period of time that was possible before the thing deteriorated or could fossilize in the open. Besides the trees, there are fish and even a whale that has been found extending through a number of "periods" when it must have been deposited all at once. There is little question that, at least sometimes, stratographic indicators are totally wrong. The K-T boundary was supposed to have been 65 Million years ago, but it appears that the timing is not very solid. Let's continue.

FUN Dating

This stands for Fluorine, Uranium, and Nitrogen level testing. Fluorine testing relies on fluorine deposits, which take the place of bone material as it fossilizes. Uranium permeated the bone and replaces calcium during the process and Nitrogen loss levels are established from the breakdown of collagens at the same time. The values of all three elements are used to determine the age of bones to over 1 million years. As time passes organic components of materials break down, depleting its nitrogen elements. Through nitrogen dating the amount of nitrogen found in an excavated item can hint the amount of time which the object has been buried, thus suggesting the time period which it came from. The burial of skeletal remains causes significant chemical changes over time, fluorine analysis is often used on two or more objects which were found within close proximity to each other in order to determine the amount of fluorine present in the remains and whether the items are from the same or similar time periods. Other elements such as Uranium are filtered into remains through the deposits which infiltrate underground. The amount of uranium present in an object is supposed to increase overtime and uranium testing can help determine the amount of time which the objects has been exposed to such deposits. Even with all three of these things tested, FUN dating is not always accurate as the chemical features in the ground differ in all environments, thus causing different rates of decay and mineral and chemical breakdown. In saying this, FUN dating is pretty accurate in determining whether two objects found in the same site are from the same time period or if their ages are within a close proximity of each other. This one seems too iffy as well.

Amino Acid Racemization Testing

This one is used to test the amount of particular amino acid residues that are still present in ancient bone material. This method can provide dating back to a little over 1 million years, but some testing after a volcanic action has shown the ages to be in error, so the extreme heat, and limitations of oxygen make this method less accurate than we need. .

Archeo-Magnetic Testing

This tests the direction and intensity of the metallic materials that were once liquid and now are solid. This has been useful in establishing dates of ancient kilns around the world. If we knew when the earth's axis was in a particular direction, we could determine the age that the kiln was fired.

Paleo-Magnetic Dating

That brings us to paleo-magnetic dating which is similar to Archeo-magnetic dating except than it checks the absolute magnetic alignments of materials on the Earth and is useful in dating objects back to about 5 million years. It seems that the Atlantic Ocean is getting larger every year and the Pacific is getting smaller. By taking a core sample of the center of the Atlantic, one can determine when each eruption of lava was cooled by approximating the expansion as a constant since the Pacific Ocean was produced [supposedly 212 Million Years ago]. By looking at the magnetic alignment of the core components, a timeline can be created that shows when the earth axis was changed to verify Archeo-Magnetic dating. The problem is that it assumes the Atlantic Ocean has been increasing at a constant rate. If the rate is slowly decreasing all Paleo-magnetic dated items are much less old including the time when the Pacific Ocean was created and the Jurassic Age was started. As this is a major marker in dating, making this time 10 percent of the believed date would alter MANY things. What we will find is that it is more like 1%.

Dendrochronology

This is simply looking at tree rings. Count the rings and know the age of the tree. Dates in excess of 5 thousand years have been possible. Of course, you need some pretty big trees. If we could find larger ones that have been cut down we could get a pretty accurate time, but you have to sit and count each year. Do not try this at home if you are thinking about dating really old things.

Obsidian Hydration Dating

Hydration tests the influx of water molecules into the crystalline structure and is used to determine the submersion time of particular artifacts back to about 700 thousand years ago. This one is pretty good, but you have to throw stuff in the water.

Varve Dating

This is testing the amount and distribution of pollen captured in glaciated areas. By looking at these details approximate dates can be determined which are used to check other dating methods and is only good for reasonably short time durations.

Planktonic Forminfera Dating

By testing the amount of plankton shells in sediment researchers can tell a lot about temperature fluctuation over time.

Pollen Level Dating

By looking at relative levels of tree pollens, a reasonable time line can also be determined, however, time periods beyond about 13 thousand years is impractical because of a lack of cross comparison of data. The graph following shows relative pollen to date comparisons.

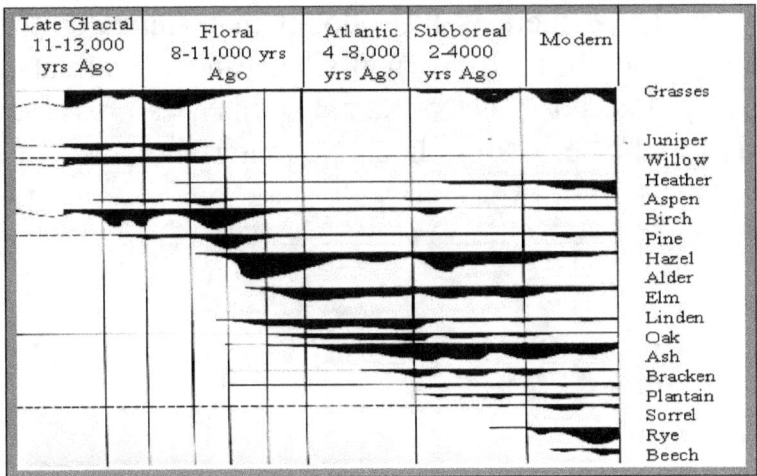

Late Glacial 11-13,000 yrs Ago	Floral 8-11,000 yrs Ago	Atlantic 4 -8,000 yrs Ago	Subboreal 2-4000 yrs Ago	Modern	

Grasses
Juniper
Willow
Heather
Aspen
Birch
Pine
Hazel
Alder
Elm
Linden
Oak
Ash
Bracken
Plantain
Sorrel
Rye
Beech

Biostraigraphy

This method uses index fossils to determine dates of other items in close proximity. Of all the methods this is certainly the most dangerous and inaccurate to use, but with other methods, it is a useful tool. It simply says if an animal was in existence during a particular time and another was near it, they lived about the same time. Unfortunately, many of the "Marker" animals were living at different times than originally were determined.

Cross-comparative Humanoid Dating

It should be noted that the ancestral humanoid finds in Africa have been dated with at least 5 separate dating methods to insure accuracy. These were Potassium-Argon, Fission Track, Paleo-magnetic, Biostratigraphic, and Carbon 14 dating. This is the type of cross comparison of data that we need to use as we try to determine the truth about our history, but we must also look at other more constant timing venues like Ice-core testing and similar things like that because nuclear timing is now being shown to be skewed. If 2 different nuclear timing methods are used and both have the same errors, the answer is no more refined. From this cross comparison, science has determined that there is little doubt that Australopithecus, probably one of modern man's ancestors lived about 2 million years ago. Unfortunately, this was characterized mostly by -the decay of radioactive materials.

.

Radio-Active Decay Timing

If the distance from the Earth to the Sun stays constant, the Sun activity is regulated, no nuclear events occur on or near earth, or things don't get abnormally hot or cold, radioactive materials decay at what seems to "generally" be a constant rate. Just count neutrons and everything goes into place, but things have not been so constant. When we say radio-active decay timing, we do not simply mean Carbon 14 dating. People have gotten so used to believing in carbon 14 markers that they don't even realize that extending the timing beyond 4 or 5 thousand years really is dangerous. Here are a few of the nuclear decay materials and methods currently being used to inaccurately time things.

Thermo-luminescence Testing

This one relies on displaced radioactive particles, which are trapped in the lattice structure of clay whenever it was "fired". By testing the alpha particle emissions, the firing date can be determined. This is especially useful if the materials have large amounts of quartz, calcite or feldspar. From the testing, dates can be easily inferred out to time periods in excess of 100 thousand years.

Extended Carbon 14 Accuracy

By passing the sample material through thermal diffusion columns for several weeks, the Carbon 14 will collect and accurate dating can now be assured out to almost 100 thousand years if no outside force changes the decay rate.

Electronic Spin Resonance Dating

This is similar to normal uranium testing except that the decay rate is modified whenever objects are buried. Testing resonance elements of the uranium by-products captures the difference. This testing provides for dating objects beyond those possible by other radiocarbon dating methods.

Fission Track Dating

Fission tracking is simply looking at the submicroscopic tracks made by uranium fission. This type of dating is useful in dating objects from 20 to 1 billion years old. Unfortunately, is as only as good as the Uranium deterioration consistency.

Ocean Sediment Testing

This method tests the oxygen 16 and oxygen 18 levels of materials in the Ocean to provide dating information.

Thorium Protactinium Dating

This is still another oceanographic dating method similar to other isotope methods. This one uses Thorium and the decomposing substance.

Argon-argon Dating

Argon-argon dating was used most frequently in dating meteorites, but rubidium-strontium and samarium-neodymium dating methods have also been used. These seem to correlate, but what they show is that the characteristics of decay are constant on a single chunk of rock floating in the universe. It does not mean the various argon and other isotope levels would have been the same on a rock on the surface of our earth during that same time, but scientists ignored this and dated the earth at 4.5 Billion years old. New evidence shows that number might be off "Significantly. The half-life for rubidium-87 is approximately 50 billion years. The decay of Sm-147 to Nd-143 has a half-life of 106 Billion years. The argon-argon method depends on the 1.25 billion year half-life of potassium-40 decaying to argon-40, which is still high. Geochronologists, the guys who specialize in telling us how old stuff is, using half-life standards have blasted out phenomenal numbers and never told us how inaccurate they could be. After all; that would limit their worth.

Lead, Lead, Lead Testing

As indicated in the previous chapter, dating the Earth and very ancient rocks is done by looking at the decay of Lead Isotopes and with 3 different types to choose from, the accuracy of this method can be very good if the decay rates would just stay the same. Radioactive decay is greatly affected by Nuclear events, Solar Flares and how close the sun is to the Earth. Most simply believe that all of these things are generally constant or none existent so nuclear decay can be considered constant. Today we know that was a mistake, but many still want to hold on to their beliefs.

Radioactive Decay Variations

Solar Flare Variation

The majority of geologists today tell you that radiometric dating has narrowed the age of Earth to about 4.5 billion years, give or take a couple of percent. We now know that is hogwash and are refining the timing more and more each day. The Earth and everything in it is much younger. Researchers at Purdue and Stanford have found evidence that radio decay rates are not constant at all. On December 13, 2006, a magnificent solar flare flung radiation and solar particles toward Earth. Measuring the decay rate of manganese-54 during the flare proved to be very interesting as the decay rate dropped during the time of the radiation fallout. It was determined that solar neutrinos zipped through space and affected Mn-54's decay rates used in the experiment. Just think about this. They were testing a single solar flare event and the change was significant. The sun has these things all the time.

Seasonal Variation

It was also found that the decay rates of silicon-32 and radium-226 showed seasonal variation, according to data collected at Brookhaven National Laboratory on Long Island and the Federal Physical and Technical Institute in Germany. This error was just the material sitting there with almost no outside interference.

Just Plane Different

Wood buried in igneous rock in Queensland Australia has been dated to 40 thousand years, while the basalt around it dated to 45 million years. Both dating subjects should have given the same date, since the igneous rock was formed at the same time the wood was buried. Many of the "data-ologists" don't tell you about major errors like this.

Lava Errors

Excess argon-36 was found in three out of 26 lava flows in recent times. So Argon/argon testing would show a much older date that actually was "KNOWN" This is believed to be because there was too much of the argon-36 in the first place. In the Grand Canyon lava flow testing showed lower levels of lava were younger than the top layers. At different volcano sites, that had eruption in 1949, 1954 and 1975. The same thing was noted These samples were dated by Geochron Laboratories of Cambridge, Massachusetts. Even though the oldest of these samples are just over sixty-years old, the lab tests provided ages that ranged from 270,000 years to 3.5 million years old. Additionally, we go to Mt. St. Helens and its eruptions in the 1980's. Samples there gave old ages in the range of 300,000 to 2.7 million years. Hopefully, you are beginning to see that we know less about how old we are than you believed before reading this.

Distance to the Sun

If neutrinos from a single solar flare can make things look older, what if the entire Earth was closer to the sun? I know that sounds odd, so just keep it in the back of your mind right now as we look at Ice Core Testing. I'll be getting back to the sun distance in a while.

Ice Core Dating

Although the task is tedious, ice can be examined just like tree rings. Each summer ice changes its consistency. $H2O(16)$ is more concentrated in the summer while $H2O(18)$ is more concentrated in the winter. This gives us indication to the level of $CO2$ which in turn allows us to understand something about the temperature levels. As the yearly cycle has freezing and thawing, ice consistency varies each day, seasonally, and yearly, depending on Earth axis and other critical elements. Anyway, scientists around the world started boring holes in ice. The most coring is done in Greenland and Antarctica. A same is shown below.

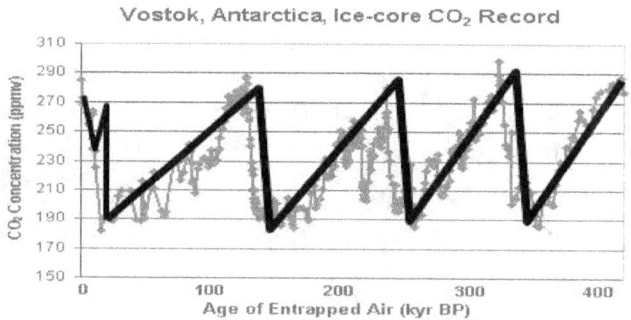

Vostok, Antarctica, Ice-core CO₂ Record

Notice, every 100 thousand or so years, there is a MASSIVE change in the $CO2$ concentration or temperature associated to some massive change. One would think this type of change would kill animals, so we might be able to use the Ice core to give us a different timeline that is better characterized by physical evidence around the world.

Before we leave this chart, please notice a sharp rise about 11 thousand years ago followed by a dip around 10 thousand years ago which indicates 2 massive climate changes occurred within a relatively short time. Yes, I know there is one of these things 220 thousand years ago, but for this history, the more recent ones are of more importance.

Greenland Check

From the next chart, we can see a correlation in near term events. 11 thousand years ago a major spike in temperature with a fast cooling followed by another just a few thousand years later then an almost flat plateau where Greenland's temperature has not changed and Greenland's position relative to the axis of spin has been unchanged. Before that time, it seems, the temperature was generally colder with what looks like a rise in temperature starting around 100 thousand years ago.

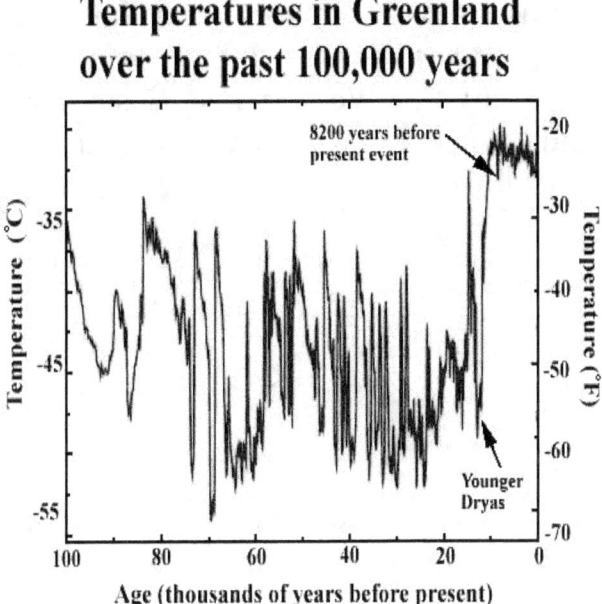

Temperatures in Greenland over the past 100,000 years

Paleo-Magnetics Again

The Atlantic Ocean is getting wider about an inch a year, averaged worldwide. While the building of the great mountains has little to do with the normal tectonic plate "drift" We can pretty accurately measure the widening ocean in various ways including measuring distances between matched magnetic landmarks on either side of a widening gap on the ocean floor. The Old theory indicated that 180 million years ago the continent Pangea began splitting apart and has been drifting ever since. In so doing, the landmasses of the Western and Eastern hemispheres separated and opened the Atlantic Ocean basin today.

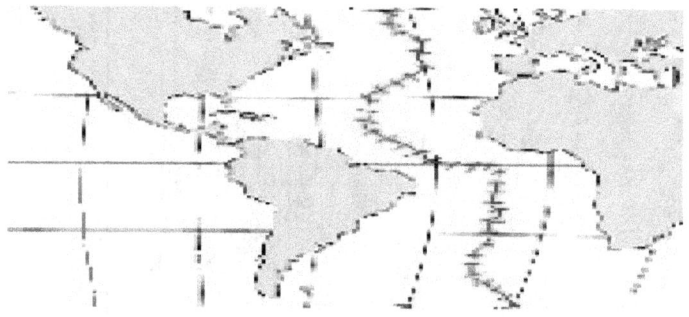

Plate tectonics tells us the outer hard crust of Earth consists actually of a dozen or so distinct, hard plates that drift individually on hot, deformable rock. An unequal distribution of heat within Earth moves the plates. The boundary between the plates forming the Atlantic Ocean is smack down the middle along the Mid-Atlantic Ridge, shown as the hashed line in the figure to the right. The ridge is where we must look to find a widening gap, which accounts for the widening ocean. That is where we measure the rate of separation.

Where the plates separate, white-hot soft mantle oozes up from great depths within the Earth to fill the gap. The molten rock cools slowly into new slivers of sea floor. This happened over and over again through the eons. That's how the Atlantic Ocean widened-by a spreading sea floor. We measure the gap rate in various ways including direct measurements of plate movement using satellite images. Another is the Paleo-magnetic method. As the Earth's magnetic poles reverse polarity periodically, the North Pole becomes the South Pole and vice versa and much of the magma spewing out is iron.

Iron-rich rock has a peculiar property: heat it above its curie point of 580 degrees Centigrade and it loses its magnetism. When it cools the rock gets re-magnetized in the direction of the existing Earth's magnetic field. So it's a magnet with the poles aligning with the poles of the Earth at the time of the cooling. The neat thing about this is: the magnetic field of the rock, once cooled, stays frozen in this orientation. It becomes a record of the Earth's field at the time of its cooling.

To measure the rate of separation, we identify two slivers of sea floor on opposite sides of the ridge that have the same magnetic polarities frozen at the same time. If you know when these reversals occur, one can simply measure the distance between magnetic alignments of the ocean floor and one can determine the rate of expansion and how long ago Pangea began to separate. Unfortunately, if the initial time-base is wrong everything is skewed.

With that, let's look at the center of the Atlantic Ocean. The graph following shows the last 14 flips over what is usually determined to be a period of 3.7 million years. These figures come from Potassium-Argon dating of magnetic material in solidified magma in the center of the Atlantic Ocean. Not only is a general time of each flip noted, but also the ferrous portions of the magma align with the magnetic field of the earth to show rotation of the earth over time. Another flip could happen any time. Using mathematical models of the external crust and inner molten material, researchers have estimated with mathematic models that the Earth should flip on its axis about every 100 thousand years. The problem with trying to determine the actual workings of the Earth is that no one has ever seen the inside of the Earth to model it properly, but the results do confirm the high possibility of a polar flip, which will cause mass destruction, tidal waves, and major climatic changes. With that scary introduction, let's look at the chart as it currently has been determined and understand that nuclear decay dating is not nearly as accurate as we once thought.

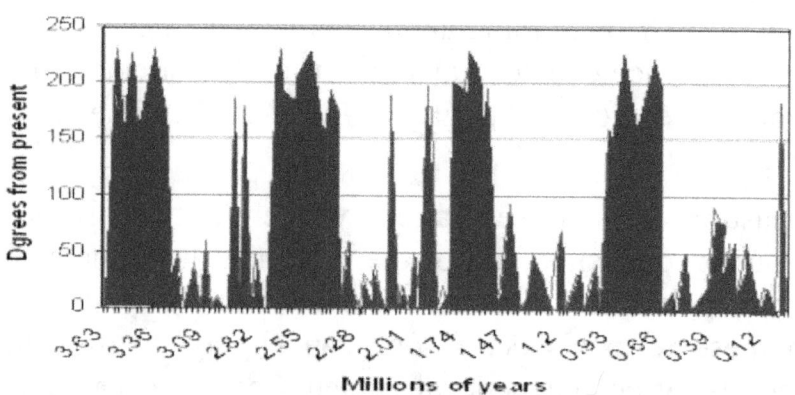

By this it shows massive shifts occurring 10, 11, 660, 930, 1470, 1740, 2260, 2550, and 3360 thousand years. They don't exactly match any of the "standard" extinction periods and they don't line up with Ice Core samples, but if we compress the timeline, look at what we find!!

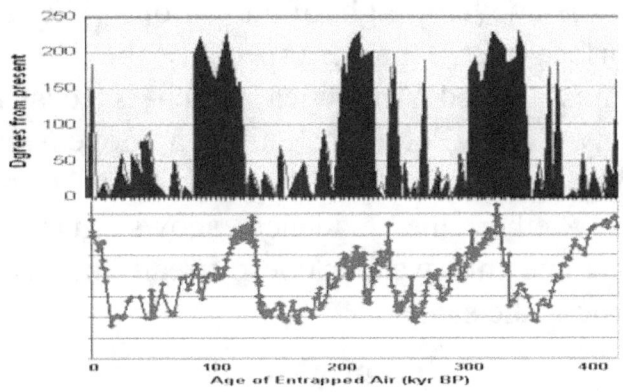

Changes in the earth axis seem to correlate very well with the data from the Ice core testing when the data is compressed. I know I haven't given you a real good reason to compress the data, but you certainly should recognize that the old data was substantially unreliable. Here is what you should recognize. The magnetic field reversals and the cyclic ice core CO_2 levels seem to have a repetitive, cyclic nature. Even that strange change around 230 thousand years ago seems to correlate with the mid Atlantic data.

I need you to notice one more thing. The compressed timing gives us more substantiation for 2 major climactic events occurring within only a very short period of time around 10 thousand and 11 thousand years ago. Later I will describe details of these important markers that can be used to help reduce the predecessor timelines down to the more valid ones I am presenting here.

Plate Shifts

Like the magnetic shifts, major crust movements or earth axis positional changes have been estimated to happen about every 20,000 years. The most recent ones occurred 43,000, 22,000, and 10,000 years ago. Sometimes the crust and magnetic field seems to wander over a number of years and other times it seems to jerk suddenly. One of the theories is that these "jerks in the crust are apparently caused by the uneven weight of the various plates supported on the surface of the Earth; especially the 19 quadrillion tons of mass called Antarctica which is located at the present day South Pole. Each time a movement occurs, terrible things happen like tropical areas turning into glaciers. Whether the evidence shows magnetic field wander or plate shift wander doesn't really matter, because the outcome is the same.

Tropical Arctic

Researchers have found evidence that the Arctic was tropical for a short time, 100 thousand years ago, or so. They found bones of early crocodiles, turtles and fish that were all tropical and estimated the summer temperatures reached into the 90s. This could only mean that the plates shifted or the planet axis moved by a substantial amount. Finds similar to this have convinced many that the outer core of the Earth moves continually and that the movement is in jerks over time.

Tropical Antarctic

If we move to the other side of the world, we find the same thing. Swamp type dinosaur bones have been found along with remains of plants that existed before parts of Antarctica became extremely cold [the last time]. It seems the animals found would have been on earth around 100 thousand years ago according to the new timing. With that little piece of data, let's look at a very special timeline track called Hawaii. Hawaii hasn't always been where it is today. A record of its travels shows up as something called hot spots.

Hot Spot Dating

If the axis is changing, there should be some dramatic physical evidence and there is. The evidence is not only from the magnetic field alignments of molten material in the Atlantic Ocean, but also some easily seen evidence. The evidence is in the form of hot spots. The best hot spot to discuss is Hawaii. The volcanic action in Hawaii has nothing to do with the edges of the plates. The picture on the following page shows the basic outlines of the major plates and these anomalous "Hot Spots". The hot spots don't stay still. They wander, but they wander in straight lines interrupted by abrupt turns. By measuring the distance the "hot spot" travels, we can determine how long the Earth or a particular plate on the Earth stayed with a particular axis of rotation. The hot spots wander because the inner core is much denser than the outer core, and occasionally the two slip in the direction perpendicular to the axis of rotation. The reason we know the slippage is perpendicular is that it is still happening.

Plate Movement Direction

If we look at the apparent trail of the Hawaiian Islands over time as shown below, a clear path is noted and times for each abrupt change has been approximated by distance. By the way, a new hotspot has just opened 73km south of the big island showing that the plate wander direction is still in the same direction as it has been over the last 10 thousand years and its perpendicular to the Earth rotational axis.

With the distance between Midway and Hawaii known to be 1300 miles, the total distance of the hotspot track is about 4500 miles. Assuming the above timing is correct, the hotspot moves about ½ inch per year. While predecessor theories put the timing of the hotspot trail as indicated above, new data has compressed the tack to agree with all the rest of the timing without the old nuclear decay standard. This makes the movement more like 50 feet per year.

Initially this sounds inappropriate as the Atlantic only increase in size about 1 inch per year today, but the motion of the hotspot has little to do with the expansion of the Atlantic Ocean as it is characterized by the differential between the Earth inside spin and the outside spin. Additionally, we get more proof of the information locked in this unusual hotspot motion by look other places. The Hawaiian Island chain isn't the only hot spot group that shows this pattern. Look on the following graphic and see that two other hot spot wander directions in the Pacific Ocean look similar to that of the Hawaiian Island spot.

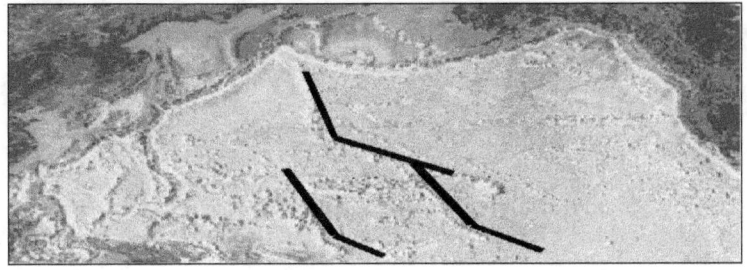

More Detail

I know you are thinking this is interesting, but it doesn't really help too much. So we have correlation of other hot spot trails and what seems to be timing compression similar to others used in this new light, but can the hot spots be traced back to the Antarctic Ice core? The answer can be seen in the next graphic. We know that the trails are produced perpendicular to the axis of rotation of the earth which is described as dotted lines below.

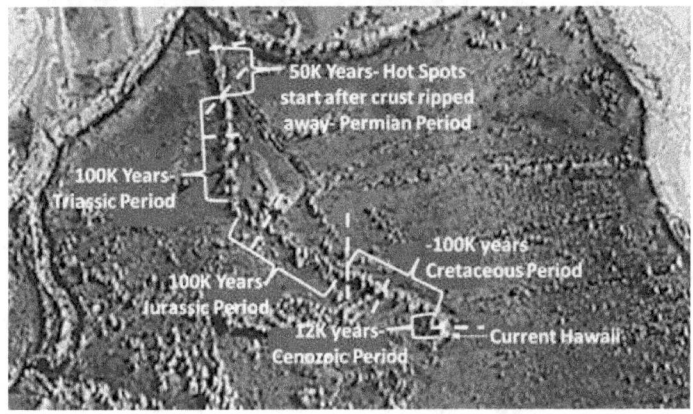

If we that the changes in the earth axis at the apparent changes, we find something VERY interesting as shown next.

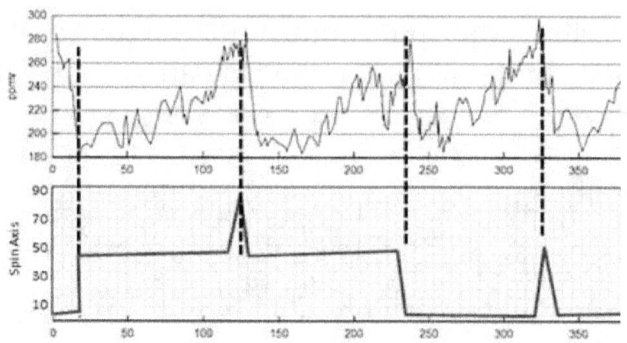

Notice that for a few thousand years about 100 thousand years ago Antarctica was probably warm between the Jurassic and Cretaceous Period. Sure enough animals from that time have been found under the ice---just sitting there waiting to be found. The graphic below tries to show some possible major earth "settling" points and general information about those spin axis's. For instance, notice that the earth spin goes along the east coast of the United States 11 thousand years ago. This will be important later as we piece all of this together to try to see critical time marks to help us reevaluate the time line for us as humans.

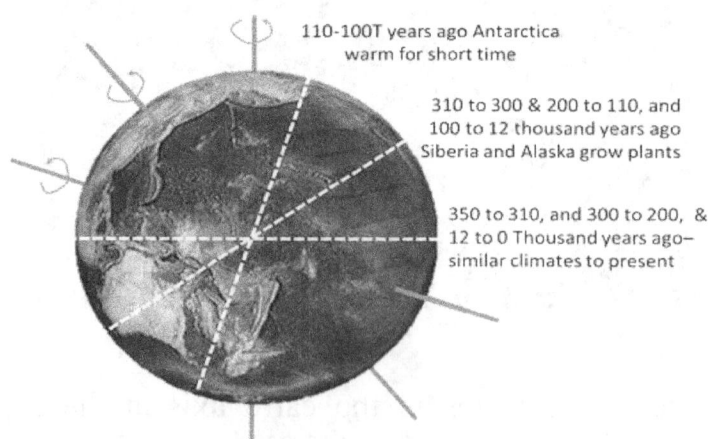

110-100T years ago Antarctica warm for short time

310 to 300 & 200 to 110, and 100 to 12 thousand years ago Siberia and Alaska grow plants

350 to 310, and 300 to 200, & 12 to 0 Thousand years ago—similar climates to present

Some Don't Believe in Shifting Poles

Some people try to infer that this whole thing about the Earth changing its axis is hogwash. Well, I think that there

42

is just way too much data to assume otherwise. Antarctica with its dinosaur bones, the quick frozen Mammoths, the various polarities of the deposited iron from volcanic action in the middle of the Atlantic Ocean; they all tell the same story. The Earth axis can move and with it there can be relatively fast and devastating climatic changes. These changes are horrible, but may not be the responsible party for most of the extinction periods. The most effective exterminator on the Earth has been and will continue to be the Comet or Meteor. Whenever a comet or meteors hit and the earth axis shifts right afterwards, total chaos occurs as it did about 100 thousand years ago then 11 thousand years ago followed by another attack 10 thousand years ago. These 3 dates are important to us as humans. A hundred thousand years ago makes the extinction of most of the dinosaurs and most of the human race at that time. After the extinction, the Bible indicated that the earth was without form and void so we can understand just how horrible it really was. Twelve thousand years ago was the last major earth axis shift and it quick froze mammoths eating in a field in Siberia when, all of a sudden, the landscape was almost immediately turned into a polar region where everything was dead. The Bible talks about this as being the destruction of the planet Rahab and other texts tell us 1/3 of the entire population of the earth was wiped out. Three thousand years later another attack temporarily shifted the earth melting the ice caps, forming massive tidal waves and drowning just about everything and everyone left on the earth. When the clamor had ended the earth shifted back to its 11 thousand year alignment as captured in the mid-Atlantic magma and the Hawaiian hot spot trail and life began again.

While all this was going on, animals would die and on special occasions, they would fossilize. For decades, scientists have been using fossilization comparisons to date things, but there were problems. One was they kept finding giant people who lived with the dinosaurs. As the dinosaurs would walk along the beaches of that day, people would come one the same beaches. We believe they went to the beach at separate times, but both sets of footprints were fossilized together, most likely before the great extinction that ended the Cretaceous Period. I'll explain the "most likely" later. Right now it is important to understand just how unstable everything is and how the nuclear decay timing could get so messed up.

The Pacific Was Made
400 thousand Years ago

Unstable Orbits

While the 100 thousand year cycle seems to be self-generated by the Earth itself, sometimes the changing characteristics that we can use as time-marks have extraterrestrial connections. The following image shows the first of three of the major ones believed to have been associated with major events involving either Mars or Venus. While many have pushed these events back hundreds of millions of years ago, from the timing corrections presented in this work, we can assume that the stability of the solar system is only recently been "standardized" into almost circular orbits around the Sun. It is the variable intensity of the sun that has caused much of the confusion in timing, so let's peel back the history just a little.

Four hundred thousand years ago the orbits of the planets weren't circular and on rare occasions, the planets came close to each other. The sun was more intense during this time period and pushed out many neutrinos into the earth's atmosphere making more and more nuclear isotopes rather than allowing them to "decay" as many had previously believed. Hundreds of thousands of years appeared to be hundreds of millions of years because of this massive illusion. The earth during this time was filled with 2 great continents. The first will be known as Pangea, but the second one, on the opposite side of the earth has no name. I'm tempted to call it Prestonia, at least while we address this unknown mass in this book. Without the second continent, counter balance, the earth would not have been stable. With all the close collisions going on, the Pacific Ocean was made. Mars got too close and pulled Prestonia right out of the earth and then Venus had its turn later.

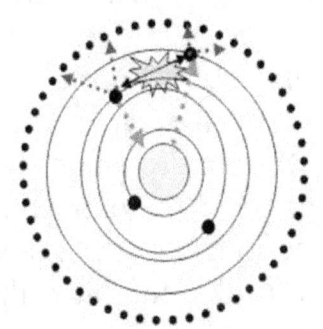 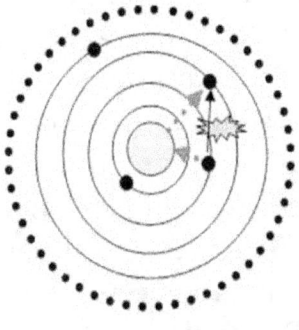

400 thousand years ago Mars almost hits earth-both planets have massive damage- Mars orbit pushed out, Earth orbit come closer to sun- Moon and Asteroids formed

12,000 YEARS AGO- close encounter Venus and Earth. Venus moved closer to sun, Earth farther away. Venus moon destroyed

Mars Mess

From the preceding diagram, notice the 4th planet which should be named Mars-plus as it is much larger than today's planet, comes near the earth that is larger as well. As it passes, the Erath crust is pulled upward into the sky creating the Himalayan Mountain Ridge. We can believe there was massive destruction as the earth environment was rocked by the intrusion. This event triggered the extinction between the Triassic and Jurassic periods on earth and millions of animals were destroyed. I like to think the animals on Prestonia were more evolved than those on Pangea, but we may never know. One thing we believe about this time period is that humans lived and walked on the land. Not only were these people living on the earth, they were civilized to some level and they were giants according to almost every ancient text known. The Bible called this group the "great men of old", but I really liked the name from the Greek. They called them the Titans. The image below shows the two planets coming together

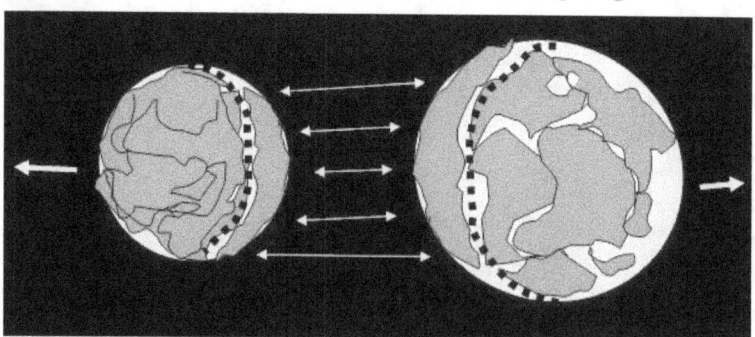

The gravitational pulls of both planets tug on the planet exterior surfaces. Along the equatorial regions of both planets massive mountain ranges pull up and perforates the exposed land. About half of the Martian surface was yanked

away and well over 1/3 of the Earth surface pulls away as well. A portion gets caught up as a satellite we call the moon, but most mixes together and is swept out in space until the sun's gravitational pull begins its orbit in the area known as the Asteroid belt.

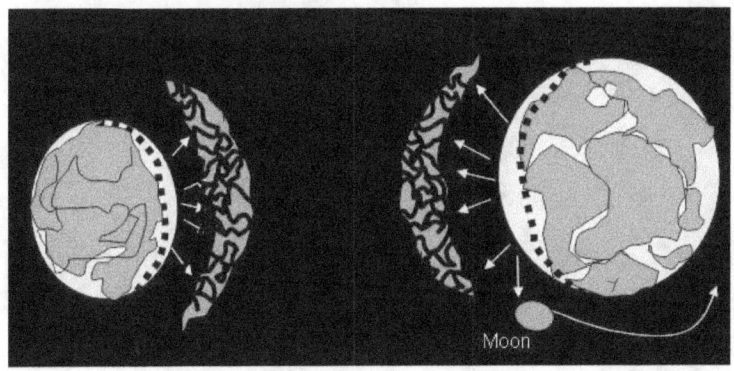

While it is very difficult to determine when all this happened, we can be pretty sure massive extinctions occurred. The most likely extinction was around 400 thousand years as what we typically call the Permian Triassic Boundary. Almost all life became extinct and the earth seems to have changed drastically at that time period. Even the cyclic nature of the Ice core samples show something very interesting as shown in the following graphic.

The chart below shows an extension of the ice core data from Antarctica presented earlier. What we see is that after 400 thousand years ago, there are very distinct and abrupt thermal changes every 100 thousand years. The cyclic nature continues before that time, but the events are greatly softened showing the characteristics of the Earth were different before that fateful time. Possibly this would be something about the larger planet and a smaller portion of the planet core being shifted as the earth spins.

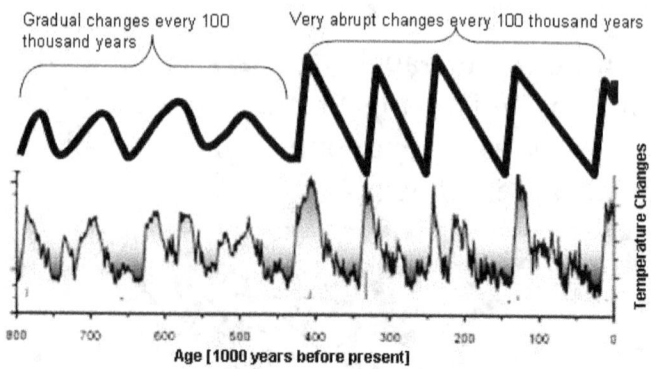

Gradual changes every 100 thousand years

Very abrupt changes every 100 thousand years

Temperature Changes

Age [1000 years before present]

Stupid Bode's Law

Speaking of information, the diagram of our solar system that showed Mars and another planet needs to be explained before someone tries to put it into one of our Bode's Law kind of definitions. We may get insight from the ancient Zoroastrians. Here is what they had to say about the formation of the mountains.

"As the evil spirit rushed in, the earth shook, and the substance of mountains was created in the earth. First, Mount Alburz arose; afterwards, the other ranges of mountains (kofaniha) of the middle of the earth (arose).." [Some outside force created mountains around the middle of the Earth.]

Thanks to many satellite pictures and computer models, we now know that something terribly bad happened to Earth and Mars and there is a high probability that they happened at or near the same time. According to computer models, Mars and Earth had several near collisions and we are going to discuss what happened when they came together. By the way, this is not some hair-brained concept that I pulled out of my head. Many reputable scientists are currently doing a lot of research in this area. These scientists first looked at the meteor evidence on Mars.

Weird Martian Craters

Mars has a very unusual pattern of craters. Ninety percent of all the craters are on one side of the planet and I bet no one ever told you that odd fact and even though you have probably seen photographs of Mars, it didn't seem strange until now. The crater side of Mars, as shown below, contains almost all of the 2700 major craters that are over 20 "miles" across are located on this "bad" side.

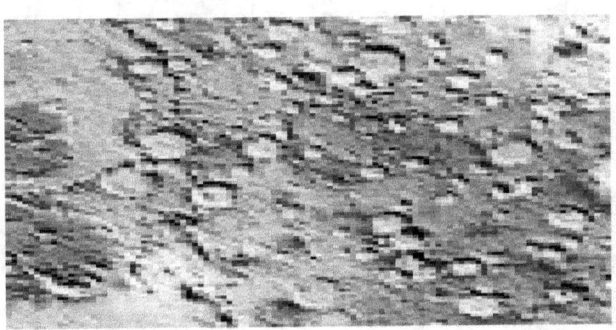

There are only two major ways that the strange cratering could have occurred. One way would be that a large planetary object got too close to Mars and exploded which, in turn, peppered the surface towards the explosion and left the other side unharmed. The mass would have had to be very close or the pieces would have been deposited around the whole planet as it rotated. The second way is similar to the first except that when the object got close to Mars, Mars itself split into two pieces, The cratered side we see today is what the whole planet looked like before the smooth half was ripped away. The second choice is the one that makes the most sense. Below on the left shows the side of Mars that has almost no craters. The crater side is shown on the right. Weird isn't it.

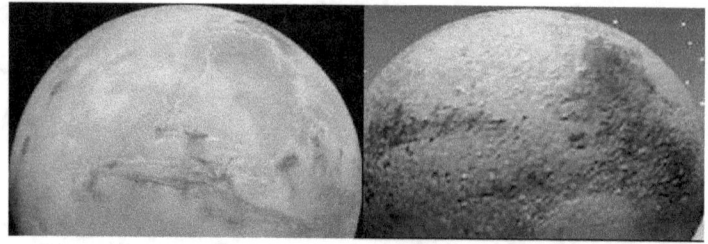

Evidence of a Split Planet

This odd cratering isn't most obvious evidence. Here is the proof from NASA. I know all of this sounds bizarre, so let's look at a topographical image of Mars. Please note that the northern hemisphere is not only smooth, but it also is sunken in much worse than our Pacific Ocean. It has a mean surface height 6 thousand meters lower than the mean of the southern hemisphere. Where do you suppose the northern half of the Planet went to?

Martian version of a Pacific Ocean

Split Mark

Also look at the dividing line between the hemispheres [I circled the area of interest]. Just like the huge slashes under the Pacific known as Mariana's trench, the northern hemisphere of Mars, is marked by one huge gash called the Valles Marineris. Although it is only 7 kilometers deep, it is up to 200 kilometers wide in places and has a total length of 4,500 kilometers. It is almost like Mars was split apart at one time and one of the marks of the split is this huge gash just like the Mariana's Trench in the Pacific.

Healing Mars

Mars, like the earth, is slowly trying to heal itself from this ancient event. In another 500 thousand years, the entire surface will probably be about the same height and here on earth, the Pacific Ocean will be about the size of the Atlantic. The drawing below shows the general topographical distinction on Mars at the present time. The dark area is the high area and it is slowly resurfacing the planet. As it does, a hole is opening at the South Pole.

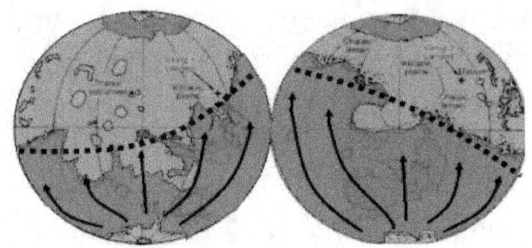

Nothing Happened?

There is a small possibly Mars has never split, but the lack of craters, low surface area and the long remaining fissure and just about everything else tells a different story.

Pulled up Mountains

On Earth we can still examine the effect of the close flybys of Mars in the form of extremely long mountain ranges. Early theories that mountains were pushed up by plates moving together did not match the positions of the mountains and people began to wonder why the mountains all fell in straight-line patterns. One path is straight along the equatorial line and another goes along the side of the Americas and along the same path on the other side of the world. The figure on the following page shows the 2 great mountain ranges. Plate tectonic models should not go around more than 50 percent of the globe, because there would be no way to push the block, but the tectonic theorists had to make them that way to support the mountain ranges until sanity finally won out.

New math models were able to capture the events that caused something quite different that mountains being pushed up. Instead, the mountain ranges had to have been pulled up. A large planetary object strafing the planet made each of the extended mountain ranges. Once the Earth was strafed with the Earth rotating on an axis through the middle of the Pacific Ocean and Asia and a second time when the rotational axis was similar to our present rotational feature. Yes, I did say the earth's axis changed. In fact, we will see that it has happened more than once. The picture on the next page shows the projected path of the Mars close encounter on two separate fly-bys. The wide lines represent the long strings of mountains along the 2 paths. One going from the southern tip of South America through the tip of Alaska and down along the coast of the Far Eastern countries. The second, more severe uplift included the region from the Middle East through Pakistan, Tibet and China.

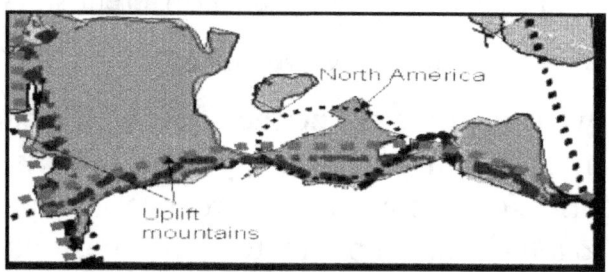

Scientific Version of the Creation of the Earth

Let's determine a "Scientific version of the Earth creation. A planet called Earth was in a widely eccentric path around the sun that put it close to the planet Mars on occasion, Its eccentric path was farther from the sun most of the time and nearer other times. Neutrinos bombardments caused less nuclear decay due to this odd feature.

- *ON Earth and probably even on Mars lived many creatures.*

- *During at least 2 of the flybys, the Earth's major mountain ranges were produced.*

- *Once, Mars came too close and a portion of the land, where the Pacific Ocean is, was pulled away from the Earth. I call this massive continent Prestonia, but most don't give it any name.*

- *Many creatures were killed on both planets.*

- *The portion of the earth that was pulled away shattered and the pieces obtained an orbit around the sun to become the Asteroid belt.*

- *Many pieces hit Mars and caused great craters on only one side of the planet. The other half of the planet had been ripped away during the encounter. It is believed that some level of atmosphere still remained, but it was not to last.*

- *The major piece of the earth was cut loose from its former orbit and took a new orbit as a new Earth.*

- *One piece from the explosion stayed with the earth and became the moon.*

- *Violent storms and floods initially filled the site of the rupture and it became the Pacific Ocean.*

- *Pangea split apart to fill the rupture and the filling is still occurring today, as the Atlantic Ocean got wider.*

Anomalies

We have all been taught about how Pangea contained all the land mass of the earth and slowly separated 200 million years ago, but there should be 3 nagging things in your head.

- *Why was Pangea the only continent?*

- *Why was the Earth damaged when Mars should have gotten the worst of it?*

- *Why does the 200 million year time period not match the Ice core samples and other newer timing components?*

Pangea Anomaly

Why was all the landmass of the earth clumped together in the form of Pangea in the first place? With all the land mass in one tiny, tiny area, the roundness of the earth would have been questionable, but the size of the earth dictates that it will establish a spherical shape. The spherical shape would have had to have been established millions of years before Pangea was formed. If we believe that Pangea existed; and there is great amounts of data to confirm its existence; we are left with an unanswered anomaly. Certainly, our students would have been taught to question the whole Pangea concept on this one anomaly alone. So, we are left with the question. How could all the land in the whole world be clumped into one place? The answer is simple.

It wasn't!!!!!!!!!

I think you are angry at the lie that has been told you? I know the historians and scientists have all lied to make our history more pleasant, but that isn't what is needed.

Mars Should have Split Anomaly

Even with the above evidence many will say—Bah Humbug—to this whole concept. The near collision of Mars and Earth must be false because physical law requires that the smaller planet would have sustained the greatest damage. As we have already explained, THEY ARE RIGHT; at least in their theory. Let's look at the remains of Mars for a minute. I already mentioned that only half of the planets has significant amounts of cratering, but what I possibly didn't emphasize is the fact that while the earth split open during the flyby, Mars got the worst of it. It essentially split in two. That fact is so obvious, it is almost comical.

Part of Mars is just like the Pacific Ocean on earth. There is almost no crustal mass remaining after the split. Unlike earth this is over the ENTIRE northern hemisphere of Mars. The crust is rarely more than a few kilometers thick and the sparsely cratered surface is suggestive of a relatively new surface. Like the remaining portion of the earth, not including the Pacific Ocean, the southern hemisphere of Mars has a strikingly thick crust, which exceeds 20 kilometers in places, and a much more heavily cratered surface. It is in this hemisphere that we find nearly all the major impact basins such as Hellas, Isidis and Argyre with crater basins well over 1 thousand Kilometers in diameter. These huge holes were probably made by some of the large chunks of earth that left during the explosion.

200 Million Year Old Anomaly

The last does not have a conclusive answer, but it seems that the Ice core timing and other elements MUST be used instead of the favored nuclear decay. Without a standard for nuclear decay, we are left with the elements brought out in this book which indicates that the last Martian flyby occurred about 400 thousand years ago rather than millions of years ago. This should give us concern about the instability of our Solar System, but it does not mean the answer is wrong.

The 200 Million Year Possibility

Don't get me wrong about this. Even if the 200 million year timing for the scooping out of the Pacific is correct, it does not change the other time line. To be complete, let me provide you with the elements that point towards the 200 million year old date. According to data collected from the Deep Sea Drilling Project [1968 to 1983] the Pacific Ocean Basin was determined to be youthful. While the youthful adjective was used, they also suggested this date was approximately 200 million years old. This dating was determined by testing mantle depth and sedimentation over the ocean floor. For the 400 thousand year date to be correct, sedimentation would have to be more rapid that they believed was appropriate.

According to several models, the break-up of Pangea occurred around 200 million years ago. One simple test was the expansion of the Atlantic. The Atlantic Ocean is the major separating line of the super continent and it is getting wider by about 2 to 3 centimeters per year with a current average width of about 6,000 kilometers. Therefore, if the separation remained constant throughout this whole period, the close encounter and the explosion that made the Pacific Ocean would have occurred around 200 million years ago. The problem with this timing is that there are more stresses on the expansion of the Atlantic when the axis is where it is today. There have been many changes in the earth's axis so that is the first problem. The second is that the initial separating pressure would have been tremendously more powerful when the event had just happened which would have separated the Atlantic much, much more rapidly that its current speed.

Two major, life form-reducing events occurred around this same time. 250 million years ago 95% of all life forms became extinct and 212 million years ago 90 percent of the remaining life forms were eradicated by a second event. There is a strong possibility that one of these events was the very close encounter with Mars that caused the Pacific Ocean. The problem with this supposition is that those dates are almost totally relying on nuclear decay being constant.

The date of the Pacific bottom, the date of the Pangea split, and the date of the most extensive extinctions point to the most traumatic event of our earth's history--the encounter with Mars. The timing of this event is not well understood. What we do know is that the Pacific Ocean was scooped out and many animals become extinct. Somehow, some survive. One of the survivors or a new type of being is this gigantic human the Greeks called the Titans and the Bible called the Great men of old. Most of the ancient texts of the world simply called this time the Age of the Giants. From ancient histories we are told that Titans soon became civilized and lived in close union with this place called "Heaven". After tens of thousands of years, their society and science had broken free from simple to unbelievable. They began warring, built war machines, potentially had nuclear energy, had electricity, and used genetic manipulation to force the modifications of animals thousands of times faster than the calculated evolution method. Besides the things we read about, huge animals and lush forests, researchers are also finding the remains of civilized humans who lived and worked in relative peace for millions of years. Today we know that humans roamed the Earth hundreds of thousands of years ago and complex societies lived during this time. All of a sudden, all evidence of them seemed to vanish. That by itself is interesting, but these guys started fighting between each other and it even got worse. First let's look are physical evidence of their existence.

Time of the Giants
400 to 100 thousand Years ago

People Walking With Dinosaurs

This section probably is describing dinosaurs and people living together before the last Extinction period 100 thousand years ago. That period of time is enough to allow for proper fossilization of the footprints. Rock strata from Triassic, Jurassic, and Cretaceous periods contain literally billions of dinosaur tracks, and actually outnumber bones by orders of magnitude. After all, a dinosaurs only made one skeleton and many footprints in its lifetime, so we can get a better understanding of dating from the footprints.

Tracks Found Everywhere

Dinosaur tracks have been found in over 1000 locations throughout the world, on every continent except Antarctica. In the U.S., they are especially abundant in Texas, Colorado, Utah, Arizona, New Mexico, Connecticut, Massachusetts, and New Jersey. It is believed that in the western U.S. alone new sites are being reported at the rate of about 50 per year. Most of these tracks have been found where there once was shorelines Large expanses of moist sediment were so important in building proper fossilized tracks. Here is the weird part. Human footprints are being found with the dinosaur ones. Many times these footprints show humans that were huge lived with the huge dinosaurs.

At Rocky Hill, Connecticut can be found a track floor that is covered with hundreds of theropod tracks.

In Amherst, Massachusetts one can find thousands of lower Jurassic dinosaur tracks from the Connecticut Valley of New England.

At Glen Rose, Texas one can see many large Cretaceous carnosaur and sauropod tracks still in their original positions.

At Seneca, New Mexico the site contains hundreds of ornithopod and theropod tracks..

Tuba City Arizona has a site containing many lower Jurassic theropod tracks.

In Denver Colorado several Cretaceous dinosaur trackways can be seen still in their original position.

At Alberta, Canada a vast collection of dinosaur tracks from the Peace River of British Columbia can be seen along with one of the largest exhibits of dinosaur skeletons.

In Price, Utah one can see displays that include about 50 Cretaceous dinosaur tracks collected from coal mine roofs.

On and on we could go, but the thing that is unusual is that some sites have human footprints mixed in. The graphic below sows some of the trackways.

Ancient World of Giants

The massive animals of the past were matched with massive people. I need to get into some of the details of this group as they lived between the time that the Pacific Ocean was created after Mars came too close and they began disappearing as a new entity was defined in many religions around the world called "Watchers" or Angels. It seemed that somehow these massive people became these watchers either after they died or some other way, but they began living in a different dimensional universe we sometimes call Heaven. This book is not about all of this stuff, but I need to give you a quick overview and show some of the details to show consistency of information. Let me just give you a very brief listing of a tiny fraction of the evidence of these massive "Titans" before we go on.

I know all this sounds absurd if you haven't heard it before, but there is plenty of historical and scientific evidence of the Titans described in the Book of Genesis as the "Great" men of old. Titans didn't just live in the Middle East. They lived all over the place and we have found evidence of their existence around the world.

Turkey

In the late 1950's during road construction in Homs southeast Turkey, many tombs of Giants were unearthed. These tombs were 4 meters long, and when entered in 2 cases the human thigh bones were measured to be over 47 inches in length. It was calculated that the person [or Titan] who owned this Femur probably stood at fourteen to sixteen feet tall.

Mexico

1925-According to the Washington Post, June 22, 1925, and the New York Herald-Tribune, June 21, 1925, a mining party found skeletons measuring 10 to 12 feet, with feet 18 to 20 inches long, near Sisoguiche, Mexico. These also sound like Titans.

South Africa

In South Africa, a giant footprint of a woman measuring over 4 feet long has been dated [by those radioactive decay methods] at approximately 9 million years old. Pointing to the probability of this being a female human-like species' foot, proportionally the two-legged being would need to be some 30 feet tall!

Australia

At Inverell Australia we find an example of the Titan evidence long, long ago. The footprint below right shows the tremendous size that some of these people got. This particular one was found along with many others.

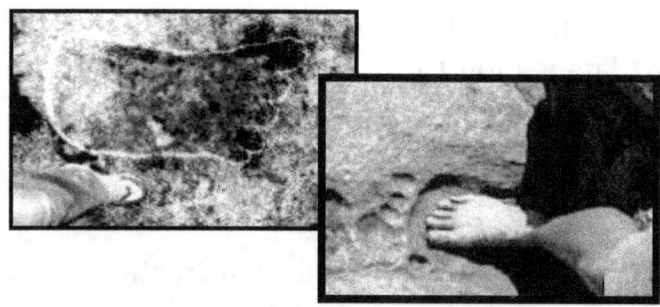

In September 1993, another giant-sized hominid fossilized footprint was added [above left]. Here, embedded in the rock, was a large footprint impression. The track was that of a right foot, probably distorted in the original soft mud, and was 44 cm in length x 30 cm across the toes. There were signs that other tracks had been embedded nearby, but these had weathered away. The imprint was preserved by a lava flow that was "reportedly" dated more than a million years ago! The monster hominid whose single footprint still survives must have been enormous, at least 11 feet in height.

Arizona

In 1923, Mr. Samuel Hubbard discovered the remains of giants in the Grand Canyon of Arizona. The discovery consisted of the following: Petrified bodies of two human beings about 18 and 15 feet in height respectively. One of these was buried under a recent rock fall which required several days' work to remove. The other, of which Mr. Hubbard took photographs, was in a crevice and inaccessible. The bodies were formed from limestone petrifaction and embedded in sandstone during a very ancient time.

Nevada

In July, 1877, four prospectors were looking for gold and silver outcroppings in a desolate, hilly area near the head of Spring Valley, not far from Eureka, Nevada. One of the men spotted something peculiar projecting from a high ledge. The prospector was surprised to find a human leg-bone and knee cap sticking out of solid rock. He and his companions dislodged the oddity with picks. Realizing they had a most unusual find, the men brought it into Eureka, where it was placed on display. The stone in which the bones were embedded was a hard, dark red quartzite, and the bones themselves were almost black with carbonization showing its great age. When the surrounding stone was carefully chipped away, the specimen was found to be composed of a leg bone broken off four inches above the knee, the knee cap and joint, the lower leg bones, and the complete bones of the foot. Several medical doctors examined the remains, and indicated that they had indeed once belonged to a human being, and a very modern-looking one. But for us the best part was their size: From knee to heel they measured 39 inches. Their owner in life had thus stood over 12 feet tall. Compounding the mystery further was the fact that the rock in which the bones were found dated to the era of the dinosaurs, the Jurassic Era. The local papers ran several stories on the marvelous find, and two museums sent investigators to see if any more of the skeleton could be located. Unfortunately, nothing else but the leg and foot existed in the rock. Again, this is just a tiny sampling of huge amounts of information.

Titanic Civilization

These titan guys were not lumbering fools as some of the Greek histories try to portray. They were very civilized and skilled in all types of science and technology. Here is a sampling of some of what has been dug up.

Building Materials and Art

Iowa-1897-A large stone [2x2x1feet] with multiple faces of an old man carved on it and a grid pattern on the remaining area was found 130 feet down in a coalmine. The estimated age was in excess of 100 million years old. [Below left]

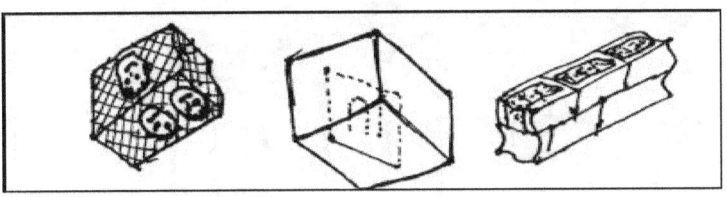

Philadelphia 1829-A 30 cubic foot piece of marble was excavated from a depth of 60 feet. Inside the marble was a straight edged rectangular indentation. After a section of the marble was carefully removed it was found that 2 distinct heavily engraved letters similar to an "I" and a "U" eleven inches long and 5.8 inches deep were on a square base. The estimated age was over 65 million years old.

Oklahoma-1928- [Above Right]A block wall was found almost 2 miles deep in a coal mine. Each block was 12 x 12 x 12 inches polished on the outside and filled with gravel on the inside- There were multiple reports over 150 yard length of the same wall. The estimated age of the wall was well over 100 million years old.

Strange Geode

The old battery in a geode-The picture below is some type of power conversion device found inside a geode, in California. Below the geode is a drawing of x-rays of the geode showing the elemental parts. These include a spring, core, plate, and electrical insulator. The same parts as you would expect in a battery. Maybe this is a new way to package batteries, but it takes a long time to complete the package. Both of the objects are extremely ancient and certainly before we originally thought that everyone used electricity. The central metal core surrounded by the white material looks like a battery. Whatever it was, it was electrical. On the right is a drawing of the parts and a size comparison to a standard D-cell battery.

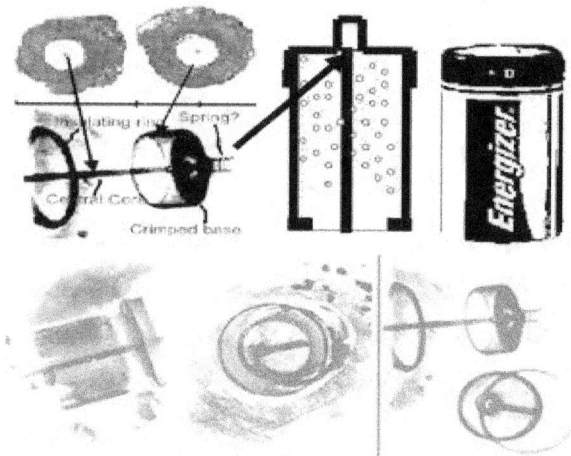

Another Geode

If that wasn't strange enough, still another geode was found with more goodies left behind by the ancient humans. Geologist Mike Walters found this geode. Inside was a worked metallic bar or ring. This could only have been man-made.

We have found many, many other items attesting to a high level of civilization of these Titans. Just think about this almost any metal object would have be completely disintegrated by now so the idea that any evidence is almost beyond belief, but there may be more.

Nuclear Material for Power and War

Electricity, metallurgy, high level manufacturing, the art of building massive walls, and the rest were nothing compared to their power to initiate war. One way to control a war is with nuclear power. In 1972, a French analyst named Bougzigues working at the Pierrelatte nuclear fuel processing plant detected a small but significant change in an important ratio between Uranium 235 and Uranium 238. The mystery uranium was eventually traced back to the Gabon mine at Oklo, Africa. You need to understand that what was so incredible was that a nuclear reaction had occurred such that plutonium was created and that the nuclear reaction itself had been moderated! [Presumably by Titans] This means that once a reaction is initiated, they were able to keep it from exploding and releasing all the energy at once. Unlike our current ones, this ancient group of reactors was, incredibly moderated using --water. Besides the obvious huge increase in sophistication with water reactor rods, you've got to really be trying to manufacture plutonium--it's a complicated process.

The picture above is the unassuming Oklo mine area with its water cooling system and the reactors. Below the image is a portion of one of 13 uranium veins that have been controlled for hundreds of thousands of years. Studies indicate that this prehistoric reactor was several miles in length. Even more astonishing is the fact that the radioactive wastes have still not migrated outside the mine site [none of that pesky nuclear waste to get rid of]. The Titan civilization was, without a doubt, technologically superior to today's civilization. So far we have found thirteen nuclear reactors existing along the 200-meter mine bed, all completely contained. We can assume, from time to time processed Uranium was take out for a number of useful things. As I mentioned before radioactive decay dating is useless around a nuclear event. In this case, the reactor was dated to be a billion years old. We now know that dating method has caused us substantial problems.

Block Growing

Block Growing by Ancient Humans was evident in the remains of their buildings. Apparently, the central core of the blocks was not as hardened as the outside rings. Layer after layer built up until the desired form was met. The shapes fit together so closely that the block must have been grown in place rather than moved to a sight fully manufactured. The pictures following have been determined to be a floor section during an ancient time in what is now West Virginia. The central core is almost completely gone, but the outside lattice structure of this ancient floor can still be seen.

In the close-up shown, note how close together each of the blocks is positioned to adjacent ones. Not even a needle could pass between their interfaces. The interior of each of the floor stones was of a softer material and eventually was eaten away.

Maryland

In Maryland we find the same pattern as the Titans lived all over North America. Again, the outside of the "grown blocks is all that remains after experiencing erosion over a hundred thousand years.

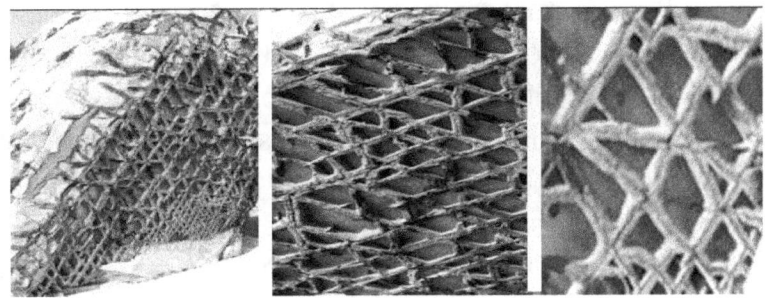

Oklahoma

The "Grown blocks" in this Oklahoma Wall shown, are believed to have been part of an ancient Phoenician metalworking plant, but they could have been much older.

If the blocks were not grown in place, where did the people find these perfectly matching stones that have not been externally shaped. We can tell the stones were not worked because the outer layer of ringed material on each block has not been violated. [See below]

I know you have never heard of a concept like growing blocks, but the evidence is pretty clear and the resulting "grown blocks" were very strong to last these thousands of years. Similar blocks have been found in Australia, Peru, Babylonia, and Egypt. This evidence suggests that these ancient humans knew things we have not yet relearned. The reason we need to understand about these Titan guys is that the timelines you were presented in the past have been so messed up; I need to expand your awareness of this time so it all falls into place. Generally, I'm still trying to let you understand how our scientists have not been completely truthful with you concerning how old things were. One of the capabilities these ancient people had was genetic manipulation and many of the changes in animal types were initiated by them. The Sumerians indicated that began building the dinosaurs to help them in the upcoming "Heaven War". I estimate this occurred about 100 thousand years ago. Around the world, ancient writers told us about this group and their genetic science. You mostly remember the Greek stories of the Titans who were on earth before the Olympian gods, but there are many more.

Ancient Persian Account

From ancient Persia we find this text that tells us the same thing.-*"Zand-Akasih"*- *Satan miscreated creatures and they became useless. God saw the defiled and bad creatures, they did not delight Him. Satan's was downfall was the unrighteous creation of the creatures and ignorance.* [Satan and the soon to be rebels during the Heaven Wars defiled God's creatures with genetic manipulation. They became unclean monsters.]

Zoroastrian Dinosaurs

This is a ninth century translation of very ancient texts from the Zoroastrian faith. In this work Ahriman is Satan, but I think you can appreciate the descriptions of dinosaurs. *"Zadspram"*-*In the beginning of creation the whole Earth was delivered over into the guardianship of the sublime. After the great rain in the beginning of the creation —after the rain, was torn up by noise and wind. A portion, moreover, as much as one-half of the whole Earth, and each of six portions around. The middle one is Pars [moon] and the lofty mountains grew up on the north so that they might become separate one from another and imperceptible. Even before the coming of the destroyer, the creatures of antagonism came to the Earth?? --And Ahriman [Satan] reported, "I will modify all types of life". They will be abominations to thee, God, and they will do my bidding." --And God said, "I will smite thee, Ahriman, and your creatures which thou thinkest have produced fame for thee. I will destroy everything about them as well." --Time made the creatures of God move differently than the Ahriman's Creatures moved. --After the noxious creatures died, and the poison there form was mixed upon the Earth*—[Satan was working with the Titans at this time. Satan's creatures were substantially different than normal creatures. Satan's animals were known as abominations or "UNCLEAN". This would have included ALL Dinosaurs. The monsters that were created by the rebel angels and Titans to fight in the heaven wars almost all died in the wars. This happened 100 thousand years ago. Afterwards, Dinosaurs were still being made.]

Hittites Genetic Manipulation

In this book, the Hittites had a somewhat funny version of the genetic manipulation in our ancient past. *"Story of Anus"-During a plot to overthrow the Taru, Kumarbi lay with a Rock as if it were a woman. The rock bore a god that was solid rock, named Ullikummis. He was born to be used as a weapon to defeat the Taru. Initially, seventy gods attacked Kumarbi and he beat them. Taru fought with the Dragon Illuyankas and his children. A mortal named Hupasiyas helped in the dragon's destruction. Sharruma had a human head on a bull's body. Taru battled the Stone-god, Ullikummis, and finally defeated him after several battles. Kumarbi was defeated by Taru and Anus and later became 'the father of all gods'.* [Possibly referencing Titans designing Monsters to help them take over heaven, it could have been just normal rock sex, or it could have been recreation of dinosaurs even after the war. I have no idea if the rock got any pleasure from the encounter. During the Heaven Wars, dragons were also created. Like the other dinosaurs and monsters, they also were defeated. It is interesting to note that the dragon is singled out as the most powerful monster in many histories. Many types of dragons seem to be different types of dinosaurs that were recreated by the preflood scientists. Another monster was made during the wars, sort of a backward Minatare. Battles against Heaven were finally won by the Heaven team, but it was not easy. Kumarbi and the others that followed him were, most likely, the ANAK who were finally kicked out of Heaven. They were kicked out of heaven only to become the rulers of Earth.]

Sumerian Description

According to *"The Epic of Creation (Enuma Elish)"*, the Sumerians tell about the same story as the Hittites.--*The Igigi, under the direction of Taimat rebelled against Enlil, and surrounded heaven. --One of the gods [Titan], Hubur, created a horned serpent, a mushussu-dragon, a lahmu-hero, an ugallu-demon, a scorpion-man, umu-demons, a fish-man, a bull-man, and others to fight in the war. Taimat made the dragon to be as a god to fight in the war.* [These Igigi genetically altered everything. Many monsters besides the Dragon were created especially for the heaven war. Certainly the dinosaurs were made. The bull-man was like the Minotaur from Greek Mythology and like the half bull half man of the Hittite history. I know this doesn't specifically say dinosaur, but you get the picture. The dragon was not just another pretty face, but was so powerful, he was like a god. We will find this identical reference in the Jewish accounts. Daniel even was able to kill one of the last dragons in the last chapter of "Daniel". Later this monster would be known as the Leviathan.]

Mongolian Genetic Manipulation

In Mongolia, the **Book of Dzyan** became their Bible. Here is what it says. *At the fourth level [of heaven], the sons were told to create their images. One third refused & two thirds obeyed. A curse was pronounced; they will be born on the fourth, suffer and cause suffering. This is the first war. There were battles fought between the Creators and the Destroyers, and battles fought for space; the seed appearing and re-appearing continuously. They slew the forms that were two- and four-faced. They fought the goatmen [Satrs], and the dog-headed men, and the men* *with fishes' bodies.* [This is not only describing the first Heaven War, but also is discussing genetic breeding like that used to create the dinosaur monsters. Even the indication that 1/3 of the angels rebel is consistent with the Biblical version of this very ancient time in history. There have been many destruction periods on the earth. The seed re-appearing idea presented here seems to refer to re-establishment of animals after each successive destruction period by genetic replacement as I have presented earlier. The outcome of the breeding was sometimes not good. The men with fish bodies are of particular interest as the Sumerians, Dogon, and Hindu all worshiped such a creature. As I presented, these part human creatures were possibly considered monsters as well.]

Ancient Brazilian "Creations"

The Mongulala tribe of Brazil was an ancient tribe that traced their ancestry back thousands of years. Let's see what they said about creating animals. *"The gods taught us the secret of man, animals, and plants. The Blood Age was the beginning of the Mongulala history. It started immediately after the Golden Age. Critical information about the events of the era was written on animal skins."* The Golden Age here is identical to the Greek Golden Age that ended with the Heaven War. The blood Age seemed to be fitting as there were horrible wars after the Golden age of the Titans.

Greek Mythology

I'm sure you recognize the Greek version. *Gaia [a Titan?], after the defeat of her Giants, created [genetically manipulated] Typhon [half serpent/ half man] to take revenge on the gods.* [This may have been a reference to the creation of the Dragon for use in the Heaven War or afterwards.]

Titans did exist and they were master scientists. The people were so advanced that they could build genetic "copies" of animals or modify same to make larger and meaner things. If you were never told this or haven't seen any of the huge amounts of evidence, or thought Adam was the first man, let me at least give some insight to this limitation even if you are having a hard time with the corrected timelines.

Genesis 1:1-2 Golden Age

The first age [day] identified in the Genesis texts was identified by the Greeks, Mongulala, and PreMaya as the Golden Age. It was the time before what we can call the great extinction. It was a time went giant dinosaurs and people roamed the Earth in relative peace. There was war between these giants according to many accounts. Finally, it is said that the Titans began to disappear and angels were all of a sudden the rage of the ancient texts. The Egyptians and Jews called them watchers, but by some type of evolutionary process some of the Titans had become something else. The first chapter of Genesis talks about them without talking about them.

Genesis 1:3-2:3 Describes the Titans

During the 6th Age a "New Man" was created according to Genesis followed by another time of relative peace others called the Silver Age. *During the 6th Age, God told the new human to "re-plenish or RE-populate" the world.* The only way someone could Re-populate anything was that there had been people on Earth before them and they had all been destroyed. Moses simply indicated that God rested during the 7th Age. In chapter 6, Moses makes reference to *the Great [or Giant] men of Old* who were before the Silver Age. These were the Titan Giants. The Silver Age was similar to the Mongulala's Blood Age.

What Was the Golden Age?

When we talk about the Titans identified in Genesis, Jeremiah 4, Isaiah 9, and other ancient texts around the world, we are not talking about from the beginning of the Earth; we are talking about some time after Mars almost hit the earth 400 thousand years ago. The amoebas and plankton and mini-animals had all been made. Amphibians had controlled the world when all of a sudden; other animals emerged including the Titans. This Golden Age would last for many thousands of years and extend over one of the worst cataclysms to be suffered by the earth. During the Jurassic period, some of the mightiest dinosaurs were created. Make no mistake; while evolution was going on in a small way, the really big changes were coming from a directed source. This directing source was not usually the creator God. Most of the changes were directed by the Titans. Later we will find out this theme does not change. The ANAK people before the worldwide flood design many animals and the ANAKIM after the flood continue this time honored profession of Genetics. Today we are trying to relearn how to build animals, but I'm not sure it will be in the best interest for humanity.

How Bad Was The Extinction?

Let's just see how bad it was after this "great Extinction. The K/T extinction affected both marine and terrestrial faunas and a wide variety of organisms. There is considerable stratigraphic evidence that marine and continental extinctions were simultaneous. All dinosaurs living at the end of the Cretaceous were killed at the K/T boundary. [Here goes!] The losers included the titanosaurs, abelisaurs, many of the basal coelurosaurs, all the toothed birds, nodosaurids, ankylosaurids, hypsilophodontids, oviraptors, tyrannosaurids, ornithomimosaurs, therizinosauroids, troodontids, dromaeosaurids, alvarezsaurids, hadrosaurids, pachycephalosaurs, basal neoceratopsians and ceratopsids. Other terrestrial vertebrates like pterosaurs and the Asiamerican marsupials also died out. That was just the land animals. The oceans were not unscathed. Marine losses were also staggering. Marine invertebrates included coccolithophorids, ammonoids, inoceramids and rudists. Marine vertebrates such as mosasaurs and plesiosaurs all died out and as many as 57% of the plants species may have become extinct as well. Life was going to start over.

Oddly Some Things Survived

Yes, the Titans died and the dinosaurs, but some of the animals did survive to this new era. It was not a total loss, but on land, nothing bigger than 50 pounds survived. Some of the survivors included insects, amphibians, turtles, snakes and lizards, crocodilians, and most toothless birds survived. Smaller mammals such as monotremes (egg-laying mammals), marsupials and small mammals having placental births also remained after K/T Extinction/ Cretaceous Period end.

Destruction Regularity

Don't worry if you disagree with the dates of these elements. I will be going through many of them as we trek through man's history and some of the time cannot be made with a great deal of accuracy. This list also does not include random Earthquakes, and volcanic actions, which are not, considered mass extermination elements. Below are some of the more cyclic destructions that we know about.

Every 100 thousand years a huge meteor splits open the Earth. The last one of that size happened about 100 thousand years ago.

Every 100 thousand years the Poles completely reverse or become unstable for a time. The last time was about 11 thousand years ago with the Atlantis incident.

Every 20 thousand years, the earth changes its rotational axis as sort of a wobble.

Every 50 thousand years we seem to have an Ice age. The last one ended 10 thousand years ago.

Every 40 thousand years the Earth's crust evidently jumps. The last jump was probably 10 thousand years ago and the entire world was flooded.

Every 100 thousand years the temperature of the Earth seems to cycle from cold to hot. The last cycle started about 11 thousand years ago.

These factors don't include the occasional star or proton cloud that reacts with the "Oort cloud" to increase meteor bombardments on Earth and cause the really big extinctions, but you can get the picture.

The Earth is not stable.

All the Titans Die

Besides Genesis, around the world the histories tell about this group of Titans, their manipulation of animals, and finally, a huge war between those living in Heaven and those living on the Earth. This war marks the end of the Titans and introduces the second part of the first verse of Genesis. Before we get into the war and the second part of verse one of Genesis let's see what a couple of the many texts say about these titans. Jeremiah and Isaiah give us a glimpse of the civilization of the Titans that would be destroyed in the Heaven Wars as the Earth was destroyed in the war

Genesis 1:1- the world "became" without form and void

Jeremiah 4:23-24-[near the end of the wars] "I beheld the Earth, and, -- all the cities [Titan cities] thereof were

broken down. [If the cities were broken down, there must have been cities before the war.]

Isaiah 9:17-21---`*Is this the man [Satan] -- made the world like a desert and overthrew its cities [Titan cities], All the kings [Titan Kings] of the nations lie in glory, each in his own tomb.* [Kings of nations and Cities shows a substantial civilization.]

Before the wars happened, apparently, the Titans somehow changed. They may have died and become angels, or they simply "evolved" into angels. There is not much detail, but all of a sudden there were millions of Angels living in our linked universe "Heaven". The main reason we know there were millions was the level of destruction that ensued by just a portion [Books of Jasher and Revelation indicates 1/3 of the angels rebelled and caused the entire world to be laid desolate.]

The most certain thing we know is that civilized Titans were on the earth during the time of the giant dinosaurs and well before the Heaven War described in the Bible. While there are hundreds of historical references to the "Heaven Wars" that finally ended the Earth of the Golden Age, here are a few additional samples.

Peruvian Titan Story

The PreInca probably had an even farther reaching society than that considered today as the Inca stronghold of Peru. From the "*Royal Commentary of the Inca*" we get a pretty good picture of the creation as determined in this section of the world. In many ways it is extremely similar to that of the Judeo-Christian history found in the Biblical texts including the existence of giant Titans.

Viracocha created the world initially it was dark Before he create the heavens and earth, he first created a race of giants. *Giants were made from stone and ruled the world. [Giants were here a long time ago and ruled the world. We will find that this concept was universal] the giants ignored the creator's wishes and did not worship him. [The giants rebelled against heaven. This also seems to be universally accepted.] Viracocha drove a giant condor across the sun* [Sounds like flying ships were used in the wars that ensued.]

Greek Titan Story

Five ages of mankind were presented in Greek Mythology. Although not exactly creations, these periods go along with those identified around the world. It is evident that the Greeks knew about the first group of people. This section comes from "Five Ages of Man"

The 1st Age-*Cronos [God] created the Golden Age. The happiest period, where people lived and died peacefully. There was no illness and no disease. The inhabitants never suffered from hardship of war or toil of the Earth. Foods were wild and plentiful. When they died they became spirits. They became the guardians of mankind.* [This is talking about the Titans. Please notice something very important as we go along. I underlined the description of the Titans becoming Angels after they died. This is important because it ties the Biblical story together as we find Satan was one of these Angel "Spirits"]

More Greek History

Greek history is typically scoffed at due to the bad press about the Greek Mythology and their discussion about Giants ruling over humans. We are now finding out that there is more truth in the Greek myths than we once believed. Here is a sampling.

"12 Children of Uranus [sky] & Gaea [Earth] became the first Titans [Cronus was the bravest]. --The giants became haughty and were imprisoned to insure they did not avenge Uranus. Terrible was the might of the first beings. They had proud thoughts and made an attack on the gods."

Time of the ANAK
100 to 7 thousand Years ago

ANAK and Anakim

In or before the Heaven War, the Titans had disappeared along with most of the dinosaurs. In order to keep things straight as we fix the timelines, we need to get a brief glimpse of the ANAK. The ANAK people once living in the universe called Heaven, their act of war against the people of Heaven proved a disaster for them. These guys may have been sort of recycled Titans, or the angels that they came from may have been a new creation, but 100 thousand years ago they were kicked out of heaven. As losing rebels they were turned back into humans, and had something called the "Light or spirit" taken away so that they would never be able to leave this place, even after death. Quite naturally, they decided they should live forever or try to get the "spirit" back. It was hard for them. The Heaven War had destroyed just about everything. Certainly, there were some deposits of DNA either stored for safe keeping or it was taken from the newly dead dinosaurs and animals from before the war. They began recreating the animals from before. This is where most of the 1st chapter of Genesis discusses plants and animals being re-created by the Elohiym [gods].

As a note Moses purposely called the creators from chapter one Elohiym and the true creator in chapter 2 the Lord of the Elohiym who created Adam as a brand new creature and he put them in a brand new Garden located somewhere near Afghanistan by all accounts.

Anyway, the ANAK people modified some of the mammals to make Homo Erectus during what Genesis called the 6th age. Later they modified this quasi-human to make Neanderthal by changing the genetic structure. Neither of these "human prototypes were as good as the original Titans or the ANAK, for that matter so the Creator God designed what we call Cro-Magnon man about 40 thousand years ago. The Bible calls this man Adam.

The ANAK tried inbreeding with the "Adamic" humans, but they, nor their descendants the ANAKIM, would get the "spirit" that they lost, we are told. Around the world the ANAK, also known as Annunaki, Anakim, Lords of ANU, "gods of Olympia" we worshipped as gods. They lived longer than the "normal people and they had ancient knowledge.

As shown on the flowing chart the ANAK, Lords of ANU, Annunaki, and Anakim have been on the earth since the Heaven War ended and lived until about 3500 years ago when, we are told, King David of the Jewish Kingdom killed the last remnants of the Anakim giants. As the most senior King of the ANAK, the one called SATAN is the one we most remember along with Zeus, and the others who forced people into accepting by doing seemingly impossible things, living unbelievably long lives, and having long heads. The first chart shows the older dating using nuclear decay dating while the second one uses the combined timing markers previously discussed which has more consistency.

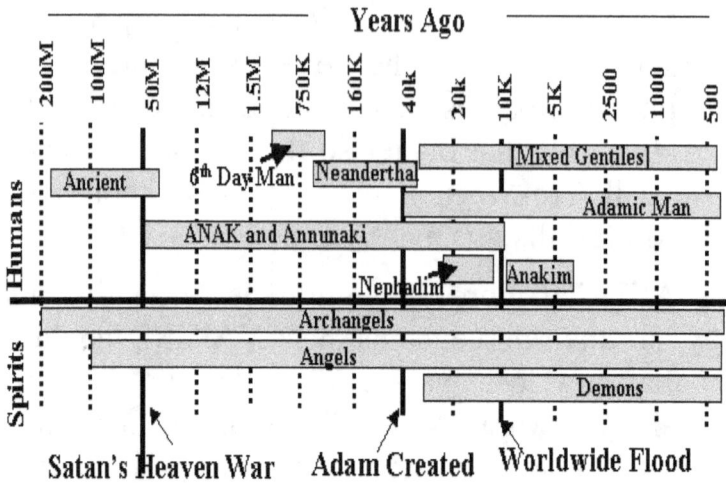

Using the updated timing method described in this book, these dates are compressed as shown below

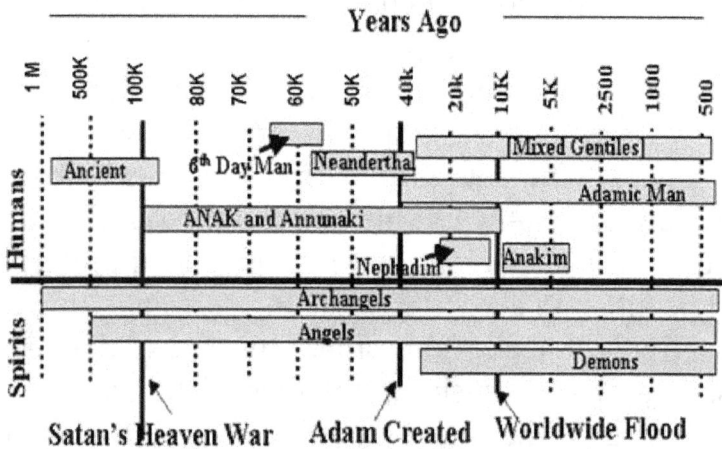

What Did They Look Like?

The ANAK we larger than most "Adamic people". While we don't know just how large the ANAK colonies were before the war, we certainly know they were huge as we are told in a number of places, 1/3 of all the people living in Heaven revolted and became ANAK. I reckon that was a huge number and it would have gotten larger as the Anakim [Descendants of the ANAK] offspring survived past the worldwide flood of Noah. We also know, generally, what they looked like. Not only were they huge people, they also had a long head, in fact the name ANAK means giant with a long head and we have found this type of skull around the world as shown below.

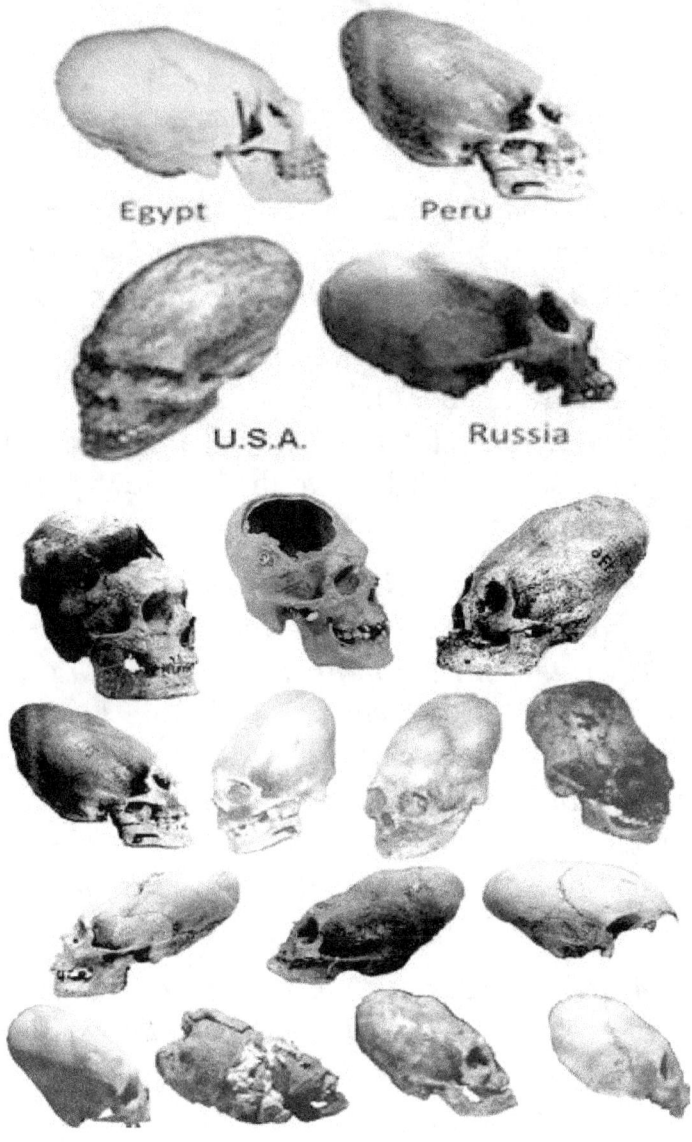

Results of the Heaven War

Genesis 1:2- *"And the Earth became without form, and void [The best way to make something desolate is to fight a war.]; and darkness was upon the face of the deep."*

This is not conclusive evidence of the war, but note how the Book of Isaiah expands this verse and others follow suit. The confirmation comes from around the world and from many religions, so don't believe that the Jews made up the whole heaven war thing.

Isaiah 14:5-7- *The LORD hath broken the staff of the wicked, and the scepter of the rulers. He who smote the people in wrath with a continual stroke, he that ruled the nations in anger, is persecuted, and none hindereth. The whole earth is at rest, and is quiet: [By this we find that there were many wars and the rulers were bad, God finally came along and wiped them all out. The whole earth was quiet. Some might say it was "Void" as the Genesis writer did.]*

Isaiah 14:8 *They break forth into singing. Yea, the fir trees rejoice at thee, and the cedars of Lebanon, saying, Since thou [Satan] art laid down, no feller is come up against us. [No one was even in the forest cutting trees.]*

Isaiah 14:9 *Hell from beneath is moved for thee to meet thee at thy coming: it stirreth up the dead for thee, even all the chief ones of the earth; it hath raised up from their* *thrones all the kings of the nations. [Again it reiterates that ALL of the kings of the Nations of the Titans were bad. This is probably saying that the kings controlled civilized kingdoms which means that many thousands of years ago society flourished.]*

Isaiah 14:10-11 *All they shall speak and say unto thee, Art thou also become weak as we? art thou become like unto us? [As a punishment Satan (and his rebels), they were turned into Normal People]*

Isaiah 14:12 *How art thou fallen from heaven, O Lucifer, son of the morning! how art thou cut down to the ground, which didst weaken the nations! [Kicked out of Heaven and cast down to earth after weakening all the Nations. Sounds like a horrible war and a terrible banishment of the giant ANAK.]*

Isaiah 14:13-15 *For thou hast said in thine heart, I will ascend into heaven, I will exalt my throne above the stars of God: I will sit also upon the mount of the congregation, in the sides of the north: I will ascend above the heights of the clouds; I will be like the most High. [Satan and his group tried to take over Heaven.]*

Isaiah 14:16-17 *Is this the man [actually ANAK] that made the earth to tremble, that did shake kingdoms; That made the world as a wilderness, and destroyed the cities thereof; that opened not the house of his prisoners? [The world was destroyed, the Earth was shaken, and all the cities were destroyed. That was a horrible war.]*

Over and over again, Isaiah wrote the same thing to reinforce the notion that this great battle occurred, the Earth and everything in it was destroyed and Satan and his rebels were expelled to Earth and became human again. The Biblical history indicates the new Angels became proud and decided to take over heaven. They were not successful. We can believe this well documented war lasted for many, many years.

We could go on and on with the written testimony of a very ancient war with the earth becoming desolate and a mastermind [Satan] and his followers [1/3 of the population of heaven] being banished to Earth. In fact, I will bring out more details so you can get an appreciation of the tremendous attempts our ancient ancestors went through to insure that we "remembered" these details. It think that the few I have selected show that everyone was told and retold about it and finally, groups wrote it down to preserve the passed down information.

Around the world, we find the same thing. A huge war between Heaven and Earth destroyer the known world, with it, the Earth became desolate. If we look at extinction periods, the last one to cause this much damage was 100 thousand years ago. As I mentioned, the Yucatan meteor hit, sprayed iridium dust around the planet, and the dust made the earth die. The problem is that they are finding most of the dinosaur bones BELOW the iridium layer so they were already dead from massive battles that had occurred prior to the massive meteor hitting the Yucatan.

Revelation of John

Revelation 12: 7-"And there was war in heaven: Michael and his angels fought against the dragon with his angels and the dragon fought and his angels, and prevailed not." [If you can't trust the visions of John, what can you trust? While this doesn't tell us when the war was, we can be pretty sure it wasn't recently.]

Egyptian Gnostic Texts

Nag Hammadi Creation Text*- and Samael [Satan] said," I have no need for anyone-it is I who am God, and there is* *no other one that exists from me"---Pisitis [God] was filled with anger and said " You are mistaken, Samael, there is an immortal man of light that has been in existence before you, and who will appear amid the creatures you have made, and will trample you, and you will descend to the abyss--- [then he and his followers] made a great war in the seven heavens.* [Rebel angels making strange animals and war that was so massive that the 7 lands in heaven are all mentioned. Now we are getting somewhere.]

Book of Abraham 3:27-*"And the second angel [good old Satan] was angry and left his estate and many followed him [Satan and the followers leaving their estate is code for rebelled against heaven] and then the Lord came down. They went down at the beginning, and they organized and deformed the heavens and Earth. The Earth, after it was deformed, was desolate, because they had not formed anything but Earth, and the spirits of the gods were brooding upon the face of the deep."* [Clearer, but not a complete picture as it appears that the earth had been made desolate so let's continue with another book.]

From Europe

Greece-Aristophanes Transcribed Myths [flowered histories]- *"Terrible was the might of the first beings [Titans that had become like gods]. They had proud thoughts and made an attack on the gods."* [Seems identical]

New Zealand Confirmation

New Zealand Maori Tradition- *"The sons of Rangi and Papa were not unanimous in the decision to separate their parents [separate heaven and Earth] so a huge war of the gods followed the separation. After a 2nd war in heaven, Tane forced rebels to other worlds [you guessed it-- earth]* *of darkness and despair."* [In this version other worlds became without form and void as the Genesis verse stated.]

Tibet Tells the Same Story

Magan Text *{Part of the "Necronomicon"}-The ancient ones that bore all the waters were one. Rebellion arose in heaven. Absu rose up to slay the Elder God, but was slain.*

Tibetan "Book of Dzyan"- This Tibetan religious document indicates that *God's sons were told to make an image, but 1/3 refused and battles were fought between the creators and the destroyers.* [Even the "1/3 of the angels" indicated in the book of Revelation is noted here.]

North American Indians

Paynut Indian Tradition-"*God Hinuno battled the other gods and some gods were thrown out of heaven."* [OK! We must remember that gods were not God, but were, instead giant humans.]

African Tribal Story

African Creation Myth-"*Initially there was a connection between God and man and commerce between the heavens and Earth. The daughter of God could visit Earth and walk with humans. The creator moved on the Earth and made more people. Something occurred to provoke a separation between heaven and Earth and death came to the world."* [The something was a huge war. Death was the same as without form and void.]

Zoroastrians

The Zoroastrian religion, which uses the *"Zadspram"* as their Bible, actually began around 700 BC. In this work Ahriman was the Angel Satan that instigated the Heaven War. After the war, the Earth was in shambles.

Zadspram *-And Ahriman was confounded and he fell back to the gloom. [Satan was defeated during the first heaven war.]---Time made the creatures of God moving, distinct from the motion of Ahriman's Creatures. [Satan's creatures were substantially different than normal creatures.One might think they looked like dinosaurs.]----After the noxious creatures died, and the poison there from was mixed upon the Earth. [The monsters that were created by the rebel angels, to fight in the heaven wars almost all died in the wars.]--- And as Ahriman came thirdly to the Earth which arrayed the whole Earth against him—since there was an animation of the Earth through the shattering. [At the end of the heaven wars the Earth was essentially shattered and was without form and void.]*

ANAK Lost the "Light"

In *"John the Evangelist"* we read about this "Light" being taken away from the ANAK- *"My father [God] changed his [Satan's] appearance because of his pride and the "light" was taken from him. His face became like a heated iron wholly like that of a man"*. [The old "heated iron man face" thing]

This thing called the "Light" runs throughout the Bible and the other ancient texts. Evidently, it is what allows someone to live in heaven. The Book of Jubilees tells us more.

Jubilees 2:9*- Nor may we take revenge on him [the Creator God] because he has stripped us of the "light". He [God] marked out the borders of the world and created man in his own image with whom he hopes again to populate heaven, with pure souls.* [The angels could not take vengeance on any of the heavenly host without this light thing. They lost some substantial power. Also note that the word "again" is put in the verse, to let us know that man was here before the war and was recreated after it was over.]

Heaven lost angels during the Heaven war with Satan, as the rebel leader. It seems that those who want to go to heaven will need this "Light" thing. It is some type of key.

ANAK Jealous of the Adamics

In the following Gnostic Works we find that Samael [Satan] wants to turn the Adamic humans into his servants.

And Samael said 'come let us create a man out of us; so that when he sees his likeness, he might be enamored by it. [This is talking about constructing a human from previous humans and animal DNA or something similar] *We shall make those who are born out of "light" our servants and their modeled form became an enclosure of the light.- they ignorantly created him-and when they finished Adam, his soul abandoned him.* [Not only did their Human not have a spirit, it also had no soul—in was more like an animal] *Then Zoe [the Spirit portion of the God trinity] sent her breath into Adam who had no soul. He began to move around.* [Continuously, we keep reading that the ANAK wanted the "light" back and they believed that inbreeding with Adamic humans would breed back in the light. Unfortunately, it never worked.]

Babylonian Genetic Manipulation

This Babylonian Version sounds very familiar. **'*Epic of Gilgamesh*"**--*Shamhat –[one of the ANAK] must take off her clothes and reveal her attractions. Do for the primitive man, as women [ANAK] do. She pulled not away, Enkidu was aroused. ---afterward- the gazelles saw Enkidu and scattered, for Enkidu had stripped--- his body was too clean [the hair was all gone]. His legs were diminished-he could not run as before, he had become wiser—Enkidu, you have become like a god.* [I assume the powerful ANAK, known as Annunaki by the Babylonians, sort of forced themselves to sleep with this lowly creature. Their desire was to make a better servant. The hybrid mutation became intelligent. By this the ANAK created something like Homo-Erectus just as the book of Genesis indicates.]

Gnostic Descriptions

Gnostic histories tell us similar accounts about the ANAK. Found in the Nag Hammadi Desert in Egypt the *"Creation Text"*, and *"Book of Abraham"* were mentioned previously, and the *"Book of Melchizedek"* fills in a couple more holes of this powerful Kingdom.

"Book of Melchizedek"- Pray for the offspring of the angels, together with seed which flowed forth from the father of all who made the entire universe from nothing there were engendered the gods [ANAK] and angels, and the men that came out of the seed, all of the natures, those in the heavens and those upon the Earth—now the nature of females was wanting among those that are in the heavens. [This is talking about the Homo erectus and Neanderthal humans before Adam] ---They were bound with men and women, but these were not the true Adam nor the true Eve. [This verse also talks about a difference between angels and creatures called gods [ANAK] and infers that a union between man and one of the ANAK was accomplished. It specifically indicates this human was not the true Adam and Eve so we are talking about 6th day human that started as Homo-Erectus and became Neanderthal].

Cretaceous Extinction

What do we know about this event? It is called the Cretaceous mass extinction of the dinosaurs. Virtually no large land animals survived. Plants were also greatly affected while tropical marine life was decimated. Global temperature was 6 to 14°C warmer than present with sea levels over 300 meters higher than current levels. At this time, the oceans flooded up to 40% of the continents. Death was everywhere. While ancient texts tell us this was the "Chalk extinction", science came along an proved its importance by finding CHALK. This event is marked by a boundary called the K-T Layer or the Cretaceous-Tertiary boundary. I know it should be C-T Layer, but I am not a scientist and didn't name the thing. Actually, the "K" is short for kreide, the German word for chalk because vast amounts of chalk were formed during the end of the Cretaceous. I know that explanation is lame, but who are you going to yell at? This K-T layer thing has been found in both marine and terrestrial sediments and at numerous boundary sites around the world. Many believe that a huge meteor hit near the Yucatan peninsula and something probably did, however, it was not necessarily a meteor. From Biblical and other texts it indicates that at the end of the Heaven War, Satan was literally thrown out of heaven and he and his minions hit the earth with a huge amount of force. Ezekiel and Revelation contain just some of the depictions. Whatever the reason for the "Chalk", this event marks the end of the age of the Titans.

The Book of "Ezekiel"

[Chapter 28] indicates that Satan was cast to the earth in the midst of "Stones of Fire" as witnessed by the kings of that very ancient day and fire consumed all things nearby. [Just imagine how angry a certain party was to throw someone to the ground and huge meteors followed him.]

The Book of "Revelation"

[Chapter 12] indicates that Satan fell to earth like a falling star and many of his followers followed him. Presumably, this was the description of the beginning of his punishment for starting the war in the first place. [You have to hit hard if you look like a falling Star. We find out that he hit so hard, the other side of the earth split open.]

Death Before Chalk

Anyway, whatever happened, the Iridium in the boundary layer can also been attributed to another source besides a meteor. The earth's core could have done it if the core somehow got to the surface. That is where India comes in. The entire country wasn't here before the K-T layer was formed. Whatever hit the earth near the Yucatan caused the earth to split open and start spewing out magma. This continued until the countries were produced and the air was filled with soot. Today, hundreds of cubic miles of magma still fill the area called the Deccan Traps. [Yes, I said MILES] By the way, dinosaurs weren't all killed by the Chalk as some have indicated. Many, many dinosaurs fossils have been found below the K-T layer showing they were killed a thousand years or so before the Chalk event.

One thing that is known by Scientists; while Iridium is only found on meteors and deep inside the earth, the amount of iridium found around the world in the K-T layer is far too dense to have come from a single meteor. The reason they don't tell you is that most of the iridium came from the

Earth and the Country of India tells the awful story. The graphic shows how chalk killed the giants.

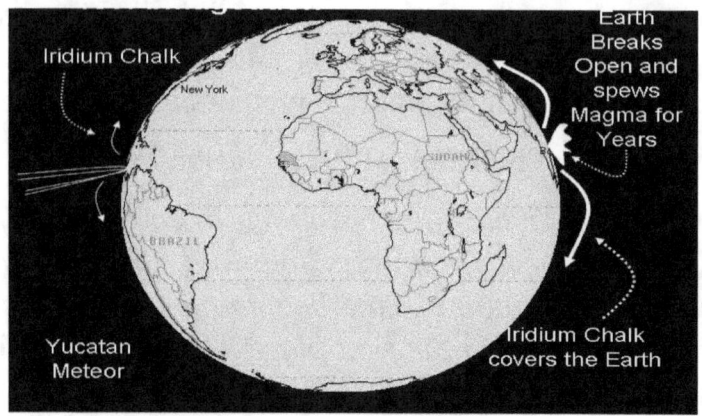

When the Yucatan "Meteor" hit, the other side of the world belched forth millions and millions of tons of Magma from a massive opening. This huge pile of magma is called India today, or technically, the Deccan Flats portion. Originally, this massive pile of "earth insides" covered over 2,000,000 square meters and contained about a million cubic kilometers of magma rich in iridium. This sort of thing has happened before man was on the Earth, but this time was the only one since mankind was created and it seems to have caused the destruction of the Titan people if any remained after the Great War period that ended the Golden Age. As the Titans disappeared and the ANAK came along to control the world, some of the timing of this time period is problematic. We need to first understand that all this fossilization is not as "standardized" as you were told in the past. Sometimes things fossilized more quickly as the sediments took over cloth, skin and bone to make us miss the boat on timing. Here are a few of the examples just to open your mind to the possibilities.

Fossilization Dating

Certainly, most fossils are extremely ancient and, generally speaking, animals of a particular type lived during a particular age. Scientists knew he processes needed to have metals infiltrate objects and transform every bit of the object into a stone. Common sense alone helps us understand the immense ages. This is simply understanding how long something takes to fossilize. There are 2 problems with this type of dating. The first is that some environments allow for very fast fossilization which destroys the "natural accuracy" associated with fossilization timing. One example is a fossilized leg in a boot shown next. The problem is that the boots are fairly modern. Fossilization could not have happened hundreds of thousands of years ago, but the leg is fossilized just the same.

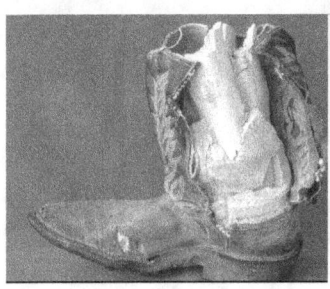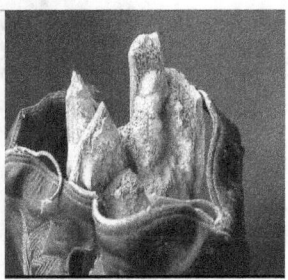

Sometimes massive mineral deposits can be injected quickly and with substantial heat buildup, the whole process can be greatly sped up. Fossilization near a volcanic site, for instance, cannot be relied on for timing.

A hat found, sort of goes with the boots, but it does not go with normal fossilization timing. It is believed that no one would make a stone hat so fossilization SOMETIMES can be done quickly as shown in the following picture. Don't get weirded out. There was no head stuck in the fossilized hat.

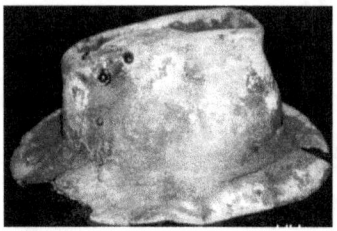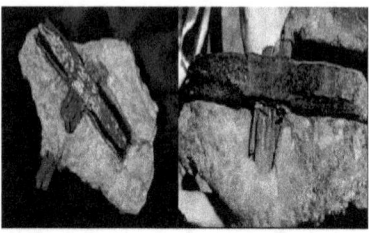

The same can be explained for the modern hammer integrated into rock shown above. There is no question the weapon is not extremely old so dating must be recalculated. Certainly, we can assume fossilization timing is not the best method of determining age.

Mazon Creek, Illinois Fossilized Child's Shoe

The following figure is of a fossilized shoe. The impression was made in ironstone, so you could not wear this thing any longer. Notice a line showing an insole and detail on the heal of the "shoe."

It is just over 4" long. The previous image of the fossilized boot could be identified as a fairly modern style, while this one could have been from hundreds of thousands of years ago. Even with an extreme date, the fossilization doesn't seem to meet the millions of years thought to be required by most Geo-scientists so dating anything takes a lot of comparative analysis, which brings us to a pretty important anomaly that plagues just about everyone trying to build a reasonable event timeline for our earth, the animals, and people. Some Dinosaurs were not fossilized. If you don't understand that genetic manipulation was a massive business prior to about 6 thousand years ago, it will mess you up every time.

Dinosaurs Were Not Fossilized

Some are now claiming that dinosaurs were not running around even hundreds of thousands of years ago. Instead, they were only very recently living on earth. This statement is both true and false. To look into this oddness, we need to look at soft tissue.

Soft Tissue Not Fossilized

Sometimes the mineral deposits are working fine, but there is a true anomaly until an answer is presented. These oddities indicate there are many examples that clearly show there is an error between what you have been told about fossil dating and what is actually more real. As I mentioned briefly before, they have found soft tissue [un-fossilized stuff] inside some of the dinosaur bones. While the geologists were squabbling, there has been pretty much consensus that soft tissues must be less than 50 thousand years old. Most say if the insides are less than 50 thousand years old the dinosaur that died must have lived less than 50 thousand years ago.

Young Triceratops and Hadrosaur

Certainly, some or most of the Cretaceous dinosaurs are extremely old, but when Triceratops and Hadrosaur bones from Montana were tested for Carbon 14 two different dating labs both said that the triceratops registered an average of 31,000 radiocarbon years and 23,000 years was the date for the Hadrosaur. That, by itself should have moved the ending date of the Triassic period up millions of years, but there were more discoveries. The picture below shows the soft tissue in a femur bone of a triceratops.

Young T-Rex

Some researchers cut open a number of T-Rex bones and found that they still had soft material that had not fossilized yet. They found collagen, elastin, laminin and other proteins showing the animal was probably less than 50,000 years old. The scientific community went berserk. This was a lie; this was a mistake; this was an anomaly. Soon all the back pointing was of no use as more and more finds showed the same thing. Some T-Rexes lived well past the time of the reported extinction.

Young Smilodon

Recent radiocarbon dating done on collagen that was taken from " femur bones" of twelve extinct saber tooth tigers, from the LeBrea Tar Pits of Los Angeles California ranged from 12,650 to 28,000 years before the present. Oops!!!

I thought everything died during the great extinction!!!

There is no question that the K-T chalk barrier marks a massive extinction. The T-Rex, Triceratops, Hadrosaur, and possibly even the Smilodon, would have been on the list of extinction receivers. The new evidence is almost irrefutable, so something happened? It is my belief and the strong indication given by ancient texts that these animals were manufactured before the great extinction period that marks the Heaven War period in our timeline. -----Then---- they were re-made well after the end of the Cretaceous period. We know the scientists of that time were well ahead of us in many technologies including genetic manipulation. It would have been fairly easy to reestablish any of the dinosaurs desired if DNA had been stored or if viable DNA was found on dead animals of that time. Sure enough, we find a number of the massive beast fossilized bones below the K-T chalk barrier marking the end of the Cretaceous period and then we find them popping up tens of thousands of years later as is event by the soft tissue, many fairly recent sightings, ancient figures of these massive beasts and radioactive bones. We even find out that they were re-created again AFTER the worldwide flood time, 7 thousand years ago.

Dinosaurs Remade

Some will come up to you and claim the earth is only a couple thousand years old because some of the dinosaurs have been found that were not completely fossilized. If it weren't for hundreds of other documents, one might even believe it after seeing the dinosaur evidence, but let's look closely. There are countless writings on the regeneration of animals into something the Bible called "Unclean" or Abominable" animals. I have picked out a few to reinforce this substantial event that has greatly mystified geo-historians, religious historians, and just about everyone else. The details are pretty clear as you will see.

Cro-Magnon Man

We all have some level of understanding about the last creation of humans called the descendants of Adam in the Book of Genesis. Some have pushed the date of his creation to something like 4000BC, but science has dated the beginning of what is called Cro-Magnon to about 40 thousand years ago. Subsequent dates for the Giza Pyramids at 40 thousand years ago and a compilation of Egyptian, Jewish, and Babylonian king lines, once normalized, showing a similar date, and many, many more pointers cue us into a strong position for the 40 thousand year beginning.

ANAK Remade Dinosaurs

Between that time and the flood of Noah was a massive increase in advancement not because Adam and Eve partook of the Fruit of the tree of knowledge, but because of the ANAK so I had better describe to you who these people were.

The ANAK were the surviving rebels from what the Bible calls the Heaven War which happened just before the Earth became void and without form as stated in the second part of the first chapter of Genesis along with many, many other places. Anyway! The ANAK, as I brought out before, were banished to the Earth but they had ancient knowledge and it was "inappropriately passed on to the "New Race". The knowledge included the working of Iron, Building of war machines, Modifying animals, and even flight. Here are some of the examples of what was said about the making of animals [Dinosaurs in particular for this book]

Enoch Genetics

7:5-6 And the ANAK began to sin against birds, and beasts, and reptiles, and fish, and to devour one another's flesh, and drink the blood.

Eve's Child Makes Dinosaurs

Eve had a number of children---60 to be exact. You know some of them, but let me introduce one brought out in the ancient Jewish book "Generations of Adam and Eve" . Her name was Ammah and she passed on her "SKILL" to her son Tranter.

[6:1-5] Among our little ones was Timnor and Ammah. Timnor understood physical law and created mighty machines. Ammah understood the secrets of creation. She manipulated the very fountain of life until she had created new forms of beings dedicated to the destruction of mankind [8:4] Timnor and Ammah practiced every abomination. Tranter learned the ways of his mother Ammah and he did manipulate the very nature of man and beast to create new forms which God had not ordained

We can imagine Ammah's laboratory. Petrie dishes filled with chromosome masses and all types of animals being constructed. THESE WERE NOT CREATIONS--- they were "Modifications"

More of the Adamics Make Dinosaurs

Enoch 10:10-11-Restore the Earth, which the ANAK have corrupted; and announce life to it, that I may revive it. All the sons of men shall not perish in consequence of every secret, by which the Angels have destroyed, and which they have taught their offspring. All the Earth has been corrupted by the effects of the teachings of Azazyal [This Azazyal was one of the people who had originally been an angel that began teaching "later" humans how to make animals.]

Almost All the Animals Were Re-Manufactured

This one is also found in the Ethiopian Bible.- *Jubilees 4:8-And lawlessness increased on the earth and all flesh corrupted its way, alike men and cattle and beasts and birds and everything that walks on the earth -all of them corrupted their ways and their orders, ---7:3- And after this [The war] they sinned against the beasts and birds.* [There are two ways to sin against beasts- sex and genetic manipulation. God didn't like either. The corrupted animals were known as "unclean" monsters. All reptiles were in this Abominable animal list.]

More References to Genetic Manipulation

This one is also found in the Ethiopian Bible and referenced in our current Bible.-*Jasher 4:18-19 and the sons of men in those days took from the cattle of the Earth, the beasts of the field and the fowls of the air, and taught the mixture of animals of one species with the other, in order therewith to provoke the Lord; and God saw the whole Earth and it was corrupt, for all flesh had corrupted its ways upon Earth, all men and all animals. And the Lord said, I will blot out man that I created from the face of the earth, yea from man to the birds of the air, together with cattle and beasts that are in the field for I repent that I made them.* [Corrupted animals did not mean the animals were evil, it meant that the species were changed inappropriately. Most had become "unclean" or "abominable" monsters. The animals were to be destroyed because they had been "modified or re-animated".]

Gnostic Jew Reference

This is one on the Gnostic book found in the Nag Hammadi Desert of Egypt many years ago. The Gnostic Teachings tell us the same thing. – ***Book of Naphtali 1:25-26-*** *The Gentiles went astray, and forsook the Lord and changed their order, and obeyed stocks and stones, spirits of deceit—become not as Sodom, which changed the order of nature.* [Not only does it indicate that the watchers or angels and gentiles practiced genetic manipulation, but also that the practice was the major cause for the flood. If you wondered why Sodom went up in smoke in the Bible, this verse indicates they were changing the order of nature-----beginning to change animals.]

More Gnostic References

ENOCH II 59:5-6- *But whosoever kills a beast without wounds, kills his own soul and defiles his flesh. And he who does any beast any injury whatsoever, in secret, it is evil practice, and he defiles his own soul.* [The killing and injury done in secret was not killing animals for food, it wasn't torturing an animal either. Neither of those things would defile one's soul. It was, most likely, genetic manipulation and corruption by integrating man's genetic material. Imagine a half human dinosaur monster.]

Still More Animal Making

This is another Gnostic Version. ***"Book of Creation"***- *Samael [Satan] said," I have no need for anyone-it is I who am God, and there is no other one that exists from me"---Pisitis [God] was filled with anger and said " You are mistaken, Samael, there is an immortal man of light that has been in existence before you, and who will appear amid the creatures you have made, and will trample you, and you will descend to the abyss--- then he and his followers made a great war in the seven heavens.* [Like the Sumerian version, Satan and his cohorts designed animal monsters to fight in another Heaven War. Many of the monsters he designed were huge dinosaurs.]

"Book of Secrets" Dinosaur Making

I saved the last 2 for now because they explain how mad God got at all the genetic manipulating and breeding and dinosaur monsters and the whole bit. The "Book of Secrets" found with the Dead Sea Scrolls provides a strong warning about the use of "secrets of God". The book simply says that if we use genetic manipulation and magic, the same thing will happen to mankind that had happened before. The earth would be destroyed again. This destruction would not be by direct intervention of God, but because we, as humans, don't understand what we are doing as we manipulate "Nature". Of the secret elements indicated in the text, it seems that the "manipulation of creation" or genetic manipulation is the worst. This seems to reference both genetic manipulation and transmutation of one material into another [Alchemy]. By all accounts, the Titans, later ANAK and humans continuously employed both of these things before and immediately after the flood. Here are the major elements of what has been pieced together of the "Book of Secrets"- judge for yourself. If it makes you fearful, you read it correctly.

Those who would penetrate the origins of knowledge, along with those who hold fast to the wonderful mysteries of magic and life- This is talking about Titans, ANAK, and "regular" humans that practice the secrets of "life". The concept of penetrating the ORIGINS of knowledge lets us know that this is very ancient science being discussed in these verses so we can believe this was done during the time of the dinosaurs to create the beasts.

With this I beseech your attention. All of the secrets of sin magic, and manipulating life were known but they [the ancient humans] did not know the secret of the way things are nor did they understand the things of old. This section indicates that no one knew the ramifications of meddling with nature before the flood. It is saying no one understood the REAL effects of using things like genetic mutation, creation of monsters, and Alchemy.

They did not know what would come upon them, so they did not rescue themselves without the secret of the way things are. Magic did not warn or save ANAK, TITAN, or Adamics from the flood. This thing called the "SECRET OF THE WAY THINGS ARE" is a secret that was not known to anyone. The monsters created were abominations.

What shall we call man who will call no one on earth wise or righteous? It is not a human possession to act on wisdom. It is not possible because wisdom is hidden except for the wisdom of cunning evil, and the schemes of Belial who modified creation, a thing that ought never to be done again, except by the command of his Maker. This is the important part. Only God has the wisdom to modify creation. Belial (a term for Satan) modified creation and it should NEVER be done again. This seems to include genetic manipulation, alchemy, and all the other magic areas.

You have not become wise in understanding my secrets of life and the earth; for you have not properly understood the origin of Wisdom. Here is all one has to do to be able to correctly use "magic". In order to understand how to manipulate nature in a good way, you must understand how it came into being. Unfortunately, you cannot because you didn't exist when this Wisdom" was established.

"The Book of Giants"
Creations

Like the Book of Secrets, the Book of Giants confirms the art of genetic manipulation and the mess that occurs from it. The best way to describe what was in the "Book of Giants" is to present the portion that is still extant. You will see that God was certainly annoyed at the giants eating men before Noah's Flood time, but according to this book and several other texts, the real kicker was the biological experimentation that the Giants did to the animals. We will never know the extent of their modification of species, but God hated it. Here is the "Book of Giants". After each verse is a short interpretation. [They are my interpretations, so take them for what they are worth.]

For they [ANAK,] knew the secrets of heaven and sin was great in the Earth because of their experiments. They made mistakes and they killed many animals and people. They had sex with women and they begat giants. Like other groups of humans, these ANAK and their offspring experimented with genetics and made monsters.

They selected two hundred donkeys, two hundred asses, two hundred rams of the flock, two hundred goats, two hundred other beasts of the field. The ANAK performed unnatural acts [on this group], and begat giants and dragons. From every animal, and from every type of human was taken its seed for mixed sex. It strongly suggests that human and animal genes were mixed together to make nasty monsters.

After a time they defiled the animals and people and begot giants, monsters, and dragons. God saw all that they begot, and, behold, all the Earth was corrupted with their blood and by the hand of man. They were brought food which did not suffice for them and they turned on mankind. They began to get hungry and they were seeking to devour many animals and people. The people ran to a safe place but the monsters and dragons attacked it. Man's flesh was eaten by all the giants. monsters, and dragons. The monsters thought that they would be saved and they would arise after death, but it was not so because they were lacking in true knowledge of heaven and because they were abominations of the Earth. If that wasn't bad enough, the giants started eating people. The giants and even the monsters did not understand that they would not go to a heavenly place after death and I'm sure their parents told them about the marvels of this other world.

They grew corrupt and did not worship the almighty God. They were considering separation of Giants]from the angels upon [the Earth, but to no avail. In the end they will perish and die because they caused great corruption in the Earth and because they tormented the Earth. Suffice to say they will be tormented after death. The giant ANAK were doomed because they corrupted just about all animal life. The way you corrupt animals is not by getting them to smoke cigarettes---it is by changing their genetic codes.

Animal Weapon

Besides the many ancient texts depicting these animals being recreated, science has found that an animal weapon was, again, adopted for the War well after the end of the Cretaceous Period. As I brought out earlier almost all Tyrannosaurus Rex skeletons have been found to be radioactive. They are so radioactive that they must be painted with heavy lead paint so they won't be dangerous while on display. I know this could be during the first war period, but the skeletons are not being completely fossilized tells us a different story. We can suppose that Ammah, Timnor's sister/wife, was one of the beginning re-creators of dinosaurs or it could have been others. The dinosaur with its tiny arms was no more effective this time than the last, but use of nuclear weapons in this time period is well known by science and massive weapon use was not only on the Earth.

Continuous War
20 to 12 Thousand Years Ago

War and Weapons

As you can imagine, if you have powerful people trying to have people think of them as gods and trying to get the most control of Earth, soon you have a mess. I'm not going to get into the hundreds of documents and evidence of the wars and how the people of that time even began to build outposts on the nearby planets, but trust me on this. It would make the book too long and the subject could get completely lost. Soon wars were everywhere, nuclear weapons were destroying massive plots of land and millions of people were dying. The book of Jasher indicated that over 1/3 of the entire population of the world was lost in the first portion of this war era. I don't know how dinosaurs were used, but it is pretty obvious that scientists of that time were making all types of "inappropriate animals" while all this was going on.

Weapons for War

If you're really smart, you make huge weapons just like we do today. Living in the antediluvian world was not a caveman existence as some would insist. Instead, there was technology. We are told and retold, these ancient people constructed IRON weapons well before the Stone Age. Today we are finding bits and pieces of this once great capability, but it is difficult in that iron rusts completely away while stone elements remain intact. Some of the more exotic materials have survived and help us understand the truth. Here are some of the ancient texts that provide details.

Genesis 4:19-22-*And Lamech took unto him two wives: the name of the one was Adah, and the name of the other Zillah [Lilith]. --And Zillah, she also bore Tubalcain, an instructor of every artificer in brass and iron;*

Adam and Eve II-1:6- *As for Cain, when the mourning of his brother had ended, he took his sister Lebuda and married her--20:1-4- Genun [another name for Tubal-Cain identified in the Genesis story] taught the children of Cain to commit all manner of grossest wickedness. He made weapons of war from iron and murder increased in the land. Genun made musical instruments and they played where the children of Seth could here. The children of Seth prayed and God rejoiced over them greatly, but soon they relaxed their prayers and the children of Cain beseeched them to come down.*

Generations of Adam-[1-:3] *Leboa, Daughter of Tamar, devised a "Sword of Light" which penetrated the wall of defense around the city of Haner and began to drain the power from the wall*

[6:1-5] *Among our little ones was Timnor -- He also created great instruments of destruction with which his people attacked the people of Cain, driving them into the mountains and dense forests, so that they could take their land for an inheritance. Thus did the evils of private ownership of the land lead to war and destruction, in which one people would destroy another to inherit their land and their goods, and the God-given abilities of mortals were turned into instruments of death and destruction.*

[8:4]- *Gringos followed the ways of his father Timnor, for he was expert with every type of machine and created many marvelous works for the service of mankind.*

[10:2]-*The king of Canaan directed his people in erecting great barriers of power around the city of Haner, so that* *none could pass into the city-neither could any missile penetrate the forces which surrounded the city.* [The barriers of power may have been more than simple walls, but missiles were used to destroy buildings during this time.]

Flying Weapon

Certainly flying craft were used for worldwide commerce during these ancient days, but during this time of unrest, many of these flying machines were used for war. Here are a few of the descriptions.

Sumerian Flying

In ancient Sumeria, Babylonia and Chaldea, many, many depictions of flying rockets, space ships were found. These drawings are some of the many depictions. Apparently, the gods used these vehicles, and carvings were made of what the people saw. The objects that the ancients saw were drawn everywhere. [Below Left]

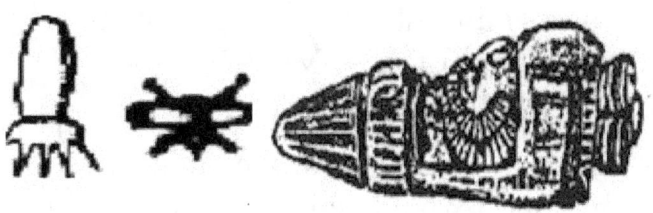

This drawing [above right] is of a Sumerian Model that was hidden away in a museum. It looks like our space shuttle, doesn't it, even down to the 3 blast engines. The model was thought to have been a very old fake for many years, however, Zecharia Sitchin examined the artifact in detail and found that it was made from a very light volcanic stone, that it was manufactured in an ancient time, certainly well before ANY thought of a space shuttle could have been imagined. It was determined that, in all likelihood, the

Sumerians as indicated by an original geological document manufactured it.

Sumerian Enlil Story- *"She valiantly ascends towards heaven. Over all the peopled lands she flies in her MU. To the heights of heaven, she joyfully wings. Over the rested places she flies in her MU."*

Sumerian Rape of Ishtar story- *[Long distance flight in one day]"One day my queen, after crossing the heavens, crossing Earth, after crossing Elam and Shubur, after crossing... the hierodule [flying ship] approached weary, and fell asleep."*

Ancient Sumerian depictions of gods flying in these vessels were clear. If a god wanted to go far, he did not fly there on his own; he used one of these Merkabas. [Above]

Sometimes humans and/or angels went with his friends in these flying machines. For those who think these are birds, I would wager that it would be impossible for a bird that size to lift three humans [Below Left]

In the middle above is a Babylonian depiction of Merkaba. This one even shows a man or angel driving the air-ship. Many carvings were made with a driver like the one shown. The Egyptian concept of the soul was a flying spirit called

Baa. The preceding graphic [right] is a picture of Baa. [Looks like a merkaba carrying an angel doesn't she?]

Chaldean and Babylonian Flying

All over the Middle Eastern world the descriptions and evidence of flying machines is overwhelming. Here are two more examples. The "Sifrala" was some kind of technical manual for the vehicles. And just like in India, being a pilot was a great honor back in the old days.

Chaldean "Sifrala"-This book provides 100 pages of technical details of how to build flying ships, with graphite rods, copper coils, crystal indicators, vibrating spheres, and something called stable angles.

Babylon "Hakatha"-This book states the following, "The privilege of operating a flying machine is great, knowledge is among the most ancient, and a gift from upon high, which is received to save lives."

Indian Flying

Indian Texts are completely filled with discussions about flying. We can believe that at least some of the flying machines were around when the others were used during the war years before the great flood. Here is a very small sampling.

India Mahabharata- *"The Gods came in their respective flying vehicles to witness the battle between Kripacarya and Arjuna. Even Indra, the Lord of Heaven, came with a special type of flying vehicle which could transport 33 divine beings."*

India Mahabharata- *"He [one of the gods] entered into the favorite divine palace of Indra and saw thousands of flying vehicles invented by the Gods lying at rest"*

Pakistan Flying

In Mohen jo Daro, drawings depicting the ruling "goddess" who was so strong she could fight two tigers at the same time. She was shown with a flying ship over her head [Some may see a spoked wheel from a chariot, but that makes no sense]. [Below left]

African Flying

In Egypt, the merkaba was well known as a flying ship. In addition to picture evidence presented previously, descriptions were written about their capabilities. To the southwest, the Dogan tribe also remembered a flying machine.

Egypt

The Egyptians filled their walls with pictures of flying ships. Some looked similar to the Australian versions shown in book one. Because the image usually means west, the image was thought to be the sun setting in the west over the dunes, but why is there sometimes 2 suns; and why is there always only ½ the sun? [Below left]

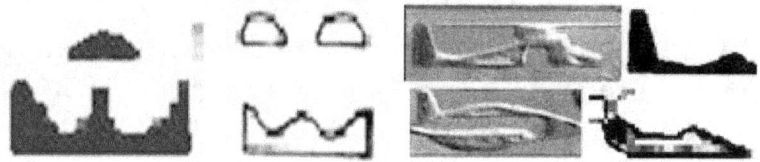

Other Egyptian drawings [Above right] looked more like the flying ships and helicopters that we have today. In fact the drawings started to show up everywhere in Egypt and that's not the only place.

Egyptian Hieroglyphics- *"Osiris and Isis [gods] descended from the sky in a sun-ship, bringing wheat, and the arts of civilization to the world."*

Egyptian "Pyramid Texts"-*"The king is a flame, moving before the wind to the end of the Earth. The King travels in air and traverses the Earth. There is brought to him a way to ascend to the sky."*

"Pyramid Texts"-*Another section of the "pyramid Texts" identifies the Pyramids as "ramps to the Sky" so that "man can go up to the heavens".* [Possibly this is referencing landing beacons.]

Egyptian "Emerald Tablets"-*"I entered the great ship of the Master. Upward we rose into the morning—[The ship was buried]---- Over the spaceship, was erected a marker in the form of a lion yet like unto a man."* [Possibly this is referencing the Sphinx. Under it there may be buried a spaceship of some kind.]

Egyptian History- *"During Zep Tepi [first time], flying gods came down to earth, flew through the air in flying boats, and gave man wisdom."*

The sketch following is of a seven inch long wooden model of an airplane found in the tomb of Padimen dated about 200 BC in Saqqara, Egypt. This model can be seen at the Egyptian museum in Cairo as it was found in 1898. Since its re-discovery in 1969, over a dozen of these models have been found. Some try to say that these are nothing more than models of birds, but what kind of bird has a vertical rudder?

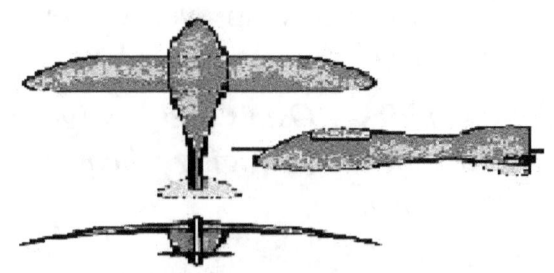

Dogan Tribal History

In their tradition, they mentioned that when the Nomos gods arrived, their ship caused a whirling dust storm as it skidded. As the ship touched the Earth, its flame went out.

Flying in the Americas

In Peru, Alaska, U.S.A., and Brazil, ships from the sky were written about and the Aztecs even provided us with a picture of one of the strange vessels.

Codex Nattal-In the Nattal, one of the pictorial images clearly shows what is believed to be a rocket ship with thruster engines and flames.

Peru Tradition-In ancient Peruvian writings, the goddess Orejona landed in a great ship from the sky.

"Popul Vuh"- In the Popul Vuh, or "Book of Council" of the Maya, we find that Hunahpu, Xbalanque, and Quetzalcoatl [all gods] returned to the stars after their life on earth ended.

Chippewa Tradition-These American Indians told of the Gin-Gwin [Flying Boats] in their historical tales.

Navaho, Pintes, and Hopi-They told of the Golden Strangers from the sky that came in flying canoes which were armed with Burning Rays. [Possibly Laser weapons]

Brazilian Manacitas Tribe-One of their cherished legends talks about the Macunbeiros, which were flying wizards that flew inside circular, luminous, machines.

Eskimo Tradition-Their tradition states that they were brought to the north by gods with metal wings.

Physical Evidence

If you don't believe the written evidence, and don't believe the pictures, will you believe physical evidence? Here are some convincing items that, at least, show that man flew in crafts similar to our airplanes and jets. Models of some of the aircraft have been found in South America. This shows that flight was not only known about, it was commonplace. This two inch long golden model of a jet-like airplane was found in Columbia, South America [Below Left]. The estimated age was 500 AD. Many other similar models have also been found. This was found in 1954.

Found around 1960

This one [Middle Above] was another of several flying models found in Columbia. This one the "historians" said was a flying fish. Two more of the flying ship models done in gold are shown above right. If those aren't enough proof there are five more below. Is there any doubt that the early Columbians saw airplanes?

Other Flying

Greek Incidents with Flying-In Greece, their early history books told of flying ships. "Flying shields" were even seen by Alexander the Great, which frightened him away from India.

Greek folklore-Archytas, one of the ancient rulers, constructed a flying machine made of wood according to ancient folklore. He also made a figure that was suspended in air by a weight from a pulley set in motion by hidden air. [Ok! I'll concede, this is a fake flying machine, but clever just the same.]

Japanese Tradition

"The spirits [angels] of the Shahman legends came to earth. Some of whom abandoned their ships at a respectable height and came down from them quietly on to Earth as if moving down an invisible ladder."

Australia, France, and China

Added to the ancient texts we find images painted on walls, models, and statues built showing what the space travelling machines were like. Next are examples of flying machines from these ancient times from Australia, France, and China. I know they look similar to those seen today called UFOs, but this was before the hype. The wars had gone into space and space vehicles were needed and used.

Australia France China

Venus Destroyed
12 Thousand Years Ago

Space War

While some scoff at the capability for having a war in space, just about every ancient text tell us over and over again about horrible battles. These battles get so bad that the great planet of Venus gets destroyed.

Psalm 148: 2-4*-Praise ye him, sun and moon: praise him, all ye stars of light. Praise him, ye heavens of heavens, and ye waters/[inhabited planets] that be above the heavens.* [Waters above the heavens is not a canopy. I would go into why no one could see the stars if a canopy was over the earth and the pressure and heat, but just know that a better explanation is inhabitable planets with water. By the way stars don't need to praise God but anyone in the stars living their might have to.]

Greek History

According to mythology, Orpheus, Apollo's son, said, *"Those innumerable souls, they fall from planet to planet and, in the abyss of space, lament the home they have forgotten."*

New Zealand

Legends indicate that *the Manu, Lofty spirits and protectors brought the Banana bush, from another star that was much further along in evolution.*

Hopi Indians

In the Book of Hopi it indicates that *there was a battle for the red city. The Kachinas, which were beings reputed to be from the fourth world came to help. In the story, some kind of tunnel was built with the speed of the wind and the Hopi were able to flee the city. The Kachinas stayed behind to defend the city and indicated to the Hopi that the time to travel back to their distant planet had not come.*

Bolivia & Brazil

Verbal tradition indicates that *the Uros beings existed before ToTiTu, the father who created the white men.* The Brazilians were "red skinned" so I'm not sure who they believed created the Brazilians, but they indicated that this Uros character came from another planet which is what I wanted to bring out here.

Wars were everywhere, 1/3 of the population of the world had already died when the fighting went into space and more died. This was a nasty war. All the ANAK leaders knew thousands of years' worth of tactics and trained a long time to control more and more of the world. Soon the nearby planets got involved. This involvement would spell the end of the war and more massive destruction than the war itself. Here are some of the many records of this expansion of war.

Babylon

In the **"Epic of Etana"** we read, *"Etana looked down and saw the Earth had become like a hill and the sea a well and so they flew for an hour and Etana looked down and the Earth was like a grinding stone and the sea like a pot. After the third hour the Earth was only a speck of dust and the sea no longer seen"* [The ship, of course, was going into outer space.]

China

Methodology of how to send a detachment of men onto any planet was described in ancient documents from Lhasa. These documents were found fairly recently and have been only partially deciphered. The remaining information is being deciphered as we speak, so we may find out more about the space war in the near future.

Greece

From Greek legends talking about battles between the gods we are told the following: *"Hot vapor lapped the titans, flames unspeakable rose bright to the upper air [outer space], lightning blinded their eyes."* [Apparently lightning, laser, or nuclear exploding/blinding weapons were used in outer space]

Ramayana, India

"Atlanteans in Vailixi flying ships" and *"Indians in Vimana flying ships" battle on Earth and Moon as recorded in the "Ramayana".*

Maharishi Bharadvay, India

In this work there are direct indications of gigantic battles in heaven.

Venus 11,000 Years Ago

This row of craters is not indicative of a meteor shower that would cause a random layout of variably sized blasts as the meteor exploded high in the atmosphere. These strikes are directed in a line on Venus. Here are seven blast areas in line. Someone was, apparently, trying to hit something during a strafing run of some kind. One thing that should be noted is each of the blast areas is exactly the same size. The blasts could not have been random pieces of meteor unless each piece came from the same source, all happened at the same time, and all pieces were the exactly the same size and density. This is indicative of bomb blasts.

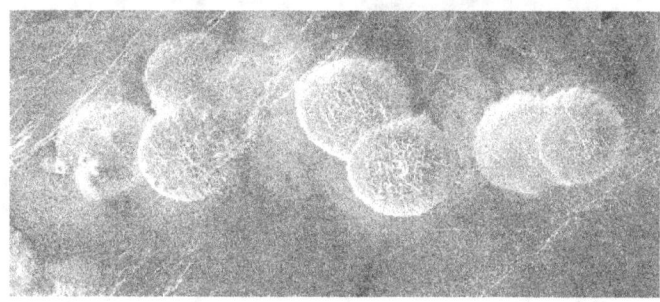

Scientists are bewildered. All of a sudden, the Pleistocene/ Holocene boundary pops up 11 thousand years ago. Massive destruction, mammoths are found quick-frozen with flowers still in their mouths, many animal types dead, meteorite holes filling the landscape, and the water level of the ocean jumped 200 feet. Additionally, we find out that Venus caught fire about this same time. Something caused the atmosphere to lose its homeostatic balance and thrust the planet into its current "cooking state". Scientists know now that the surface features on the planet associated with this unnatural heating, shows extremely young forming dates, which may be less than 50 thousand years old [I say 13 thousand year old features]. These features, could not have been caused by the "Greenhouse effect", as taught in school, without some extreme external stimuli.

Lava Frozen in Time

On Venus, the lava flows give us a strange sight. Some of the volcanoes on Venus have flows that extend out 250 miles. Below are two such lava flows. This wide explosive environment would destroy much of the landscape except that none of the volcanoes are active and the landscape stays constant. The huge volcanic actions only recently occurred, possibly as little as 12 thousand years ago, but no new eruptions have been seen. Within a short time, these lava flow indications will all melt away in the intense heat of Venus today.

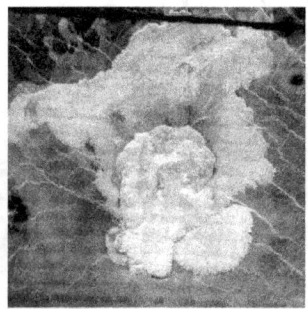 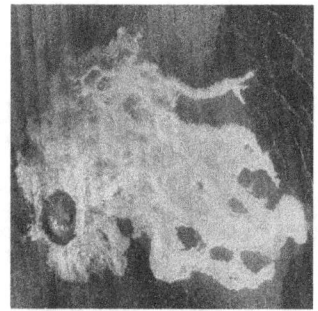

Frozen Rivers

The remains of a river delta can be clearly seen. This indicates that the destruction did not happen in the extremely distant past because the delicate tributary paths are still visible today. Below is one of the river deltas found. Water no longer flows, as it was quickly evaporated whenever the temperature rose quickly in the very recent past.

I know you have been brought up to believe that the Earth is starting to go into a "greenhouse affected state" because we used underarm spray deodorant, and the fluorocarbons are eating away our O-zone layer but the whole concept is a fabrication.

For those thinking the earth is going to catch on fire because we sprayed too much underarm spray of the Cow flatulence level was too great as some of the many government funded research tries to proclaim, you are looking in the wrong direction. Certainly looking at what happened to Venus is no cause for keeping people from putting Freon in their leaky air conditioning systems, as we have done to keep the O-zone protected. Whatever it was that initiated the burning of Venus did a great job. The entire surface and all who may have lived there perished within a very short time period. The planet became a dead land, essentially overnight, but during the process, it put on a big show that was recorded around the world.

Meteors Were Seen Everywhere

We can surmise that, as part of the wars, the inhabitants of Venus or soldiers sent to the planet, possibly, got involved and sided with one group or another and the result was not good. According to ancient records, "flames came out of the planet Venus". Pieces of the planet hit the Earth, and I mean a lot of pieces. As many meteoritic chunks came down like huge fireballs, the sun and stars were blanked out in some areas. This event was also recorded around the world, in both tradition and physical evidence. The explosion on Venus was the precursor to something like a giant tidal wave. In all likelihood this tidal wave was responsible for the sinking of many cultural outposts around the world, but before we get into this subject let's examine what the explosion caused and the memory of the event, as told by the ancients.

Did You Say, "This is Preposterous!"?

If you think all of this is preposterous, you have a number of people that will agree. Public opinion doesn't change history, or at least it shouldn't and a reasonably high probability can be assured of the Venusian meteors. We will examine events that occurred in our ancient past as recorded by many, many cultures and then examine the substantial physical evidence so you can make up your own mind.

This sounds very peculiar so some decided to see what the Bible had to say about a strange planet named RAHAB [the vain place or pure vanity]. Just to keep you from wondering, many societies call Venus the "Vain planet" so don't go thinking it was Mars or something. While it is not known the number of people that actually died in war, the book of "Jasher" provides us with estimates of the destruction during the disaster. It indicated that 1/3 of the inhabitants of the entire Earth were destroyed even before the problems with RAHAB. Here are the specific Biblical verses.

Psalm 89:10-*"Thou [God] hast broken Rahab in pieces, as one that is slain;"* [The pieces sound like meteoritic pieces. Especially as we read further. The reason Venus was destroyed was found in the book of Job.]

Job 26:12- *"The boastful Angel and his followers rebelled. Yahweh destroyed their dwelling places. He divideth the sea with his power, and by his discretion he smashed Rahab. It was reduced to stones of fire."* [By this verse we could well believe that quite a few people were living on Rahab when it was destroyed. Here is a stretch. The stones of fire could be meteors hitting the earth. The dividing the sea thing, we will find out from Plato that a lot of Island civilizations were submerged during the massive environmental changes on earth from the onslaught of meteors.]

Isaiah 51:9- *"O arm of the LORD; awake, as in the ancient days, in the generations of old. Art thou not it that hath split Rahab, and wounded the dragon?"* [The reason the dragon, (Satan and his troops), were wounded is that he was beginning to wage war beyond Earth. Note the idea that the planet being split could very well be the splitting of Venus, or destroying its moon.]

Enoch 85 and Revelation 9- *"I beheld a single star fell from heaven-then I beheld many stars which descended and projected themselves from heaven to where the first star was."* [This could very well be the vision of many meteorites hitting the Earth that scientists have found.]

Jasher 2:5-6- *"-and the sons of men forsook the Lord all the days of Enoch [Adam's great, great, great, great grandson] and his children; and the anger of the Lord was kindled on account of their works and abominations which they did in the Earth. And the Lord caused the waters of the river Gihon to overwhelm them, and he destroyed and consumed them, and he destroyed the third part of the Earth, and notwithstanding this, the sons of men did not turn from their evil ways--"* [We'll discuss how the 1/3 fits into everything shortly, but remember that the oceans began to rise and it destroyed many people. The water didn't rise as high as the worldwide flood thing, but many died just the same. By the way it, is believed that ENOCH died around 15 thousand years ago.]

Isaiah 14:12- How art thou fallen from heaven, O "Heylel" [Heylel means "morning star" or Venus], son of the morning! How art thou cut down to the ground, which didst weaken the nations! [The translation of Heylel is almost always the term "Morning Star". This is the only verse in the Bible where the phrase "Morning Star" was changed to the word "Lucifer" so that the morning star could be associated with Lucifer or Satan. It only makes sense that this is talking about parts of Venus falling to the ground and weakening the nations. The idea of weakening nations sounds like something hurting the human inhabitants of the Earth such as one would expect from the huge meteorite storm aftermath of an exploded moon of the nearby planet Venus. It should be noted, there is evidence to suggest that many of the followers of Satan were stationed on Venus prior to its explosion, so the translation could have significant merit.]

Whatever caused massive explosions on Venus is not identified, but Venus still has a gaping hole that goes 1/4 way around the planet. It turned the planet into a fireball overnight and hundreds of thousands of pieces of material became meteors that hit the earth. Some historical descriptions of these killers were recorded.

South American Venusian Meteors

The People of South America remembered and wrote about it. These Inca legends tell the story. The Inca called Venus the *"Wavy haired planet";* [This also seems difficult to interpret. Could wavy hair be flames shooting from its surface during a time when the Inca were around?]

Central American Venusian Meteors

The People of Central America remembered and wrote about it. This is from one of the Aztec and Mayan legends. The Aztecs called Venus *"the Star that smoked"* and said that it once passed by the world blazing and killing many people. The Aztec god, Quetzalcoatl, associated with Venus, is typically pictured with a wavy headdress. [I'm going to get into this whole wavy haired tail thing in a minute, but the blazing and killing sounds like meteors hitting and setting fires. The meteors came from the smoking planet.] In the Mayan Dresden Codex, the god of Venus is depicted with shooting darts. It seems to me that if something were shot away from a planet, it would have been meteor-like.

Blackfoot Indian Venusian Meteors

The People of North America remembered and wrote about it. Let's see what the Blackfoot had to say. According to their traditions, *"The morning star [Venus] put on a scarlet cloak [sounds like it turned red.] and appeared before a woman on Earth that he loved. She went into the sky with him, but was warned never to look back. She did, of course, and was ordered to return to Earth."* [The return was a mess if we believe the other histories.]

Ute Indian Venusian Meteors

The Ute Indians tell us the same thing in their verbal history. *"The sun was slivered into a thousand fragments, which fell to Earth causing a general fire. Then Ta-wats fled before the destruction he had wrought. All were consumed; until at last, swollen with heat, the eyes of the god burst and tears gushed forth in a flood which spread over the Earth and extinguished the fire."* [This flood is probably not the worldwide flood we have all heard about, but it was significant, just the same. As far as the sun bursting, I personally believe it was Venus and not the sun.]

141

Egyptian Venusian Meteors

In Egypt, the event was known and written about. Sonchie, the high priest, told Solon, a Greek historian, about events before the flood. He wrote, *"Many are the destructions of mankind that have been and shall be. The greatest are by fire and water. During long intervals there are deviations of the bodies that move around the Earth in the heavens and the consequence is widespread destruction by fire of things on the Earth."* [The fires must have been everywhere when the Venusian moon split apart. The comment that it was one of the "Normal bodies that moved around in the Earth sky" limits the body to one of the close planets. Of course, the closest is Venus.]

Jewish Venusian Meteors

The Jews wrote about the event in the book of **Enoch chapter 85 verses 1-4** we read: *"A single star fell from heaven- raised up and fed among the cows--I saw many stars which descended and projected themselves from heaven to where the first star was."* [Some claim this and similar verses are figurative and depict Satan being thrown from heaven, but sometimes people simply write what they want people to read.]

Sumerian Venusian Meteors

The Sumerians made record of the blazing tail of Venus. Their goddess named Inanna was associated with Venus and the information is the same as recorded by all the rest. *"To the queen of the heavens Inanna [Venus], to her who filled the sky with her pure blaze. The luminations are as bright as the sun. Who initiated the flood-storm? You roared in the heavens and Earth. You smote the flesh of the people."* [The blaze of Venus filled the sky, roared across Earth and smote the people. I think the only way Venus could smote the people is if its moon exploded and pieces

142

fell to Earth as a huge meteorite storm.]--- *She [Innana/Venus] who causes the heavens to rumble. She who shakes the Earthquake. She cried toward heaven and Earth, "My hair will whirl in heaven for you." You flash like lightning over the highlands. You throw firebrands across the Earth. You split apart the mountains.* [The hair extending sounds like a reference to a comet tail or a blasted away section of Venus that hit the Earth. Firebrands hitting the Earth sounds like meteors to me.]

Phoenician Venusian Meteors

Phoenician texts describe the event, but this time the goddess is Astarte, the Phoenician version of Ishtar. *"See, Astarte" [Venus], she descends into a pool as a fiery falling star".* [A beautiful description for a terrible disaster]

Persian Venusian Meteors

Mandaean Texts from Persia give us the same information. *"150 thousand years after man was created, the whole Earth broke out into flames and only 2 escaped." It continues by saying that they had children and, of those ancestors, Noh [almost like Noah] was the one that survived the Flood.* [The Earth being filled with flames could have been from the huge quantity of meteors from the explosion, but clearly this event occurred well before Noh survived a worldwide flood.]

Indian Literature

Indian literature states the following, *"Her [Venus's] anger grew so terrible that she transformed herself, grew smaller and black. On a blind rampage she was killing everything and everyone in sight. Her hair is wild, her eyes red. The world trembles and cracks under her tread. Her dark hair flies in the sky sweeping away the sun and stars."* [Again

143

we read about the comet-like tail and so many meteors that the sky is darkened.]

Assyrian Venusian Meteors

Assyrian literature tells the same story. This time the goddess is named Ishtar, but it is the same. *"To the pure flame that fills the heaven, who shines like the sun 'Ishtar"* [Venus]—*"I ran battle down like flames in the fighting. I make heaven and Earth shake. I trample the Earth. I destroy what remains of the inhabited world".* [To destroy the remains of the inhabited world, there must have been something substantial that happened with Venus.]

Arabian Venusian Meteors

Coptic texts date the event for us in the Age of Leo. The ancient Arabic text called **"Bundahishn"** tells us the following: *The Ancient Coptic text tells about a great fire and flood coming out of the constellation of Leo.* [This not only describes the event but places it in the "Age of Leo", 11 to 13 thousand years ago.] It goes farther indicating that *the beginning of world history was around 11 thousand years ago and some of the major deities were born during this event.* [The beginning of history must have meant that there was a destruction period just before that time.]

Chinese Venusian Meteors

The Far East writers also informed us of this terrible calamity. The people remembered Venus sending down a huge meteor shower. The Chinese writers said the same thing, *"There was a time when a planet [Venus] approached close to the Earth, causing great showers of stones."* [Not too many of the planets could have come close to earth. The moon of Venus is my guess.] Venus was depicted as a dangerous *"fire spitting planet"* according to Chinese legend.

144

Pacific Island Venusian Meteors

Even the people of the Pacific remembered Venus sending down a huge meteor shower. Venus was depicted as a dangerous, *"fire spitting, planet"* by the Samoans. [It is like reading the Chinese version. What would have given them that idea?]

Meteorite Evidence

Large amounts of *"meteoritic mass"* and an estimated 500 thousand strange indentions, strongly believed to be from massive meteorite showers have been found around the world that date to the end of the Pleistocene era, about 11 thousand years ago. Large quantities have been found in Alaska, Siberia, Bolivia, and Netherlands. Guess what! The time period for the destruction of the Venusian moon is about 11 thousand years ago. If they both happened about the same time, there is a good possibility that they were the same event.

Glass Evidence

Tektites are small pieces of glass formed as a meteor strikes the ground and melts the surrounding area. Many have been found in sort of an "S" shape and distributed over large portions of the Earth. Some were found embedded in fossilized wood, in Australia, others were found in Vietnam and still others were found in the Indian Ocean. Several dating methods were used including Stratographic and Carbon 14. They showed that most of these pieces were deposited around 10 thousand years ago. Ok! Maybe the ones inside the fossilized wood came from an earlier strike, but most were Pleistocene Era events just like the Venus moon blast.

New Zealand Evidence

Today, huge quantities of metallic meteorites as well as objects called "china stones" can be found everywhere on the island. Inside the stones are the remains of burned up Pleistocene type material, which dates the event to between 10 and 20 thousand years ago. [I suppose you think these came from the Venus moon strike just like me.]

Carolina Bay Evidence

The east coast of the United States was pelted with many objects. There are still an estimated 500 thousand meteorite indentions called "Carolina Bays", which mark this incredible event in history. One hundred and forty thousand of these holes have diameters of over 500 feet. Just think about how afraid the people of that time were as they essentially saw the sky fall all around them. The picture following shows the major areas where these objects have been found in the United States. These generally date around the same time. The evidence shows that the Venusian moon met its end at the same time that these 500 thousand holes appeared. Some of these indentions are very large and have diameters that are thousands of feet across. So it wasn't just a little meteorite storm. A small sampling is shown.

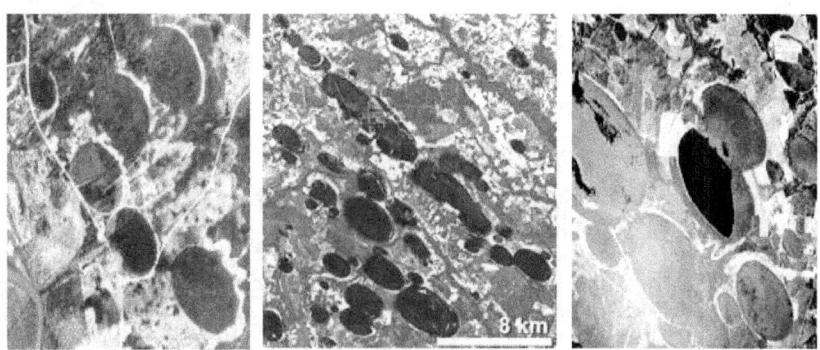

Venus Goes Closer to the Sun

It is believed that one of the reasons for the Venusian moon destruction was that the Earth came close to the planet. That would make a lot of sense in that there is a crack along the equator of Venus that goes over 1/3 the distance around the planet. This means there were huge "pull-up pressures from some massive planet that came near. This pushed Venus closer to the sun and spelled its doom forever and the other planet; earth must have stabilized its orbit to be more rounded as it is today. While this now has stabilized the nuclear decay of particles to some extent, the event only very recently happened so dates beyond 11 thousand years are greatly questionable.

The graphic following shows the 2 planets as they came in close contact. It should be known that the SOHO satellite still shows signs of plasma strings between the two planets which strongly suggest an electromagnetic discharge between the two. This indicates there was a fairly close contact at one time.

As the moon of Venus got in the way of the tremendous gravitational pulls of both planets during a close encounter, it exploded as shown.

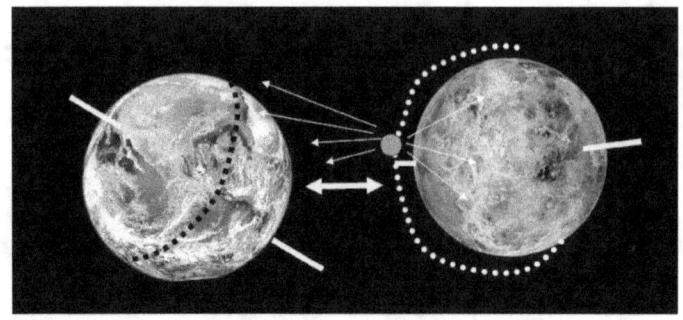

The massive fallout of the explosion riddled the earth with hundreds of thousands of meteors that finally hit along the equatorial region of that time along the Eastern coastline of North America as shown. Venus is pushed towards the sun; its atmosphere held the heat and cooked the planet.

The Earth Shifts and Floods

Dinosaurs Die Before Earth Shift

All this Venusian meteorite stuff finally shifted the earth's axis. According to the Hawaiian Island hotspot detail, this was the last major shift in the earth. I think we have a pretty good time mark for this event as I have brought out in this book. The shift was a bad one. Siberia became an ice cold wasteland as did Alaska. Egypt died up and became a desert. The shift was I know you are wondering just how we know this, but let's look at the evidence.

Sahara Was Lush

Large numbers of complete dinosaur fossils near the surface of the earth in the Sahara desert and the Sahara is not lush anymore. It would have dried up after the last shift pushed it so close to the Equator.

Alaska Was Lush

Near Alaska's Coleville River, eight separate species of dinosaurs were discovered in two sites. The included the gorgosaurus, troodon, and dromaeosaurus, edmontosaurus, pachyrhinosaurus, pachycephalosaurus, and thescelosaurus. Not only were they buried together and quickly, they must have had the food and warmth they needed to grow the plants they liked to eat and keep them warm. You guessed it the animals died BEFORE Alaska got so cold and they died QUICKLY.

Siberia Was Lush

Thousands of frozen carcasses of mammoths, some rhinoceros and musk ox have been found quick-frozen in the northern arctic latitudes of Siberia and Alaska. The farther North you go the more animals you find quick frozen in the snow. Ninety-foot plum trees which were quick frozen and over ninety feet in height with green leaves have been found in Siberia.

Wisconsin Ice Age

This is another to show what happened during this time and to help us date everything. The Wisconsin Ice Age was during this period from 20 thousand until 11 thousand years ago when it abruptly stopped. Millions of tons of Ice disappeared as the Earth shifted. Guess what happened when the Ice melted?

Many Islands Flooded

As you can imagine, the people of earth were filled with horror and massive fires swept the land and hundreds of thousands of meteors travel around the earth before blasting the eastern coast of North America. What happened next was worse and the Earth shifted on its axis as indicated by the Hawaiian track and Ice pent up in pole lands quickly melted and substantial lands sank into the ocean never to be surfaced again as the new poles were mostly on land so the water levels never went low again. The most famous sinking was of a commercial hub somewhere in the Atlantic call Atlantis but there were at least 6 more commercial sensitive Islands that were doomed by the shift from the explosions on Venus.

Ok! There are a number of underwater cities that have been found, but I'm thinking you were still of the impression that Plato made up the whole Atlantis thing as sort of a novel or something. I know this seems like a bizarre subject, but you have already gone through many bizarre subjects and you are still reading this history, so who's the oddball?

Don't get me wrong. I haven't been trying to find strange elements of conjecture to push into some type of fairy tale. I am bringing you a cross comparative collection of elements to support a probable history that just happens to link some of the stranger writings and physical evidence together with more mainstream science and religion. Atlantis fits this mold.

Everyone has heard about Plato's Atlantis description covered in the book *"Timeaus"*, but there are some aspects of his history that should be brought out a little more. They deal with lost memory, the dual flood, and the ancestors of the Greeks and Egyptians. These secondary elements tell us things about Atlantis that we need to know. In his second work on the subject *"Critius",* more evidence is provided to increase the possibility of his statements being the truth, as he knew it. The biggest thing to understand is that Plato indicated that he was told all this sinking of islands happened 11 thousand years ago when the Mammoths were being quick frozen.

Lost Memory

Plato's was explicit when describing the way and why Greeks had lost their memory of the times before the flood. According to Plato's work, the Egyptian priests told the Greek historian, Solon, that the Greeks were ignorant. According to the priest, this ignorance was "because" of a huge flood. He also indicated that the Greeks were descended from the Atlanteans. We will see later, that this "loss of memory" element not only fits with the concept of a second major flood a thousand years after Atlantis's sinking but also with the destruction of the Tower of Babel story, you think you know. This segment of *"Timeaus"* tells us emphatically that the sinking of Atlantis and at least two major floods occurred before the Tower of Babel mishap. It tells us that the first [Atlantean] sinking was very slow and the inhabitants were able to relocate in and around the area of Greece as the Island was sinking; and finally it tells us that preflood scientific and historical information was recovered and housed in Egypt. The timeline is consistent and the details have been affirmed by other historical data and physical evidence. Let's read what it actually said.

Timeaus 22d-23c-"You are all [Greeks] young in mind," came the reply: *"you have no belief rooted in old tradition and no knowledge hoary with age.* [Greeks didn't know about Atlantis sinking because something happened after the island disappearance that caused the Greek way of life to essentially have to start from scratch again.]

And the reason is this. There have been and will be many different calamities to destroy mankind, the greatest of them by fire and water, and lesser ones by countless other means. But in our temples we have preserved from earliest times a written record of any great or splendid achievement or notable event which has come to our ears whether it occurred in your part of the world or here or anywhere else; whereas with you and others, writing and the other necessities of civilization have only just been developed when the periodic scourge of the deluge descends, and spares none but the unlettered and uncultured, so that you have to begin again like children, in complete ignorance of what happened in our part of the world or in yours in early times... [This indicates that the "Atlantis sinking" was not the last major catastrophe encountered by the Egyptians and Greeks.]

You remember only one deluge, though there have been many, and you do not know that the finest and best race of men that ever existed lived in your country; you and your fellow citizens are descended from the few survivors that remained, but you know nothing about it because so many succeeding generations left no record in writing". [The most obvious reason that the Greeks had no direct knowledge of the Atlantean refugees colonizing Greece would have been that another terrible flood or cataclysm occurred a long time after the Atlantis sinking. We have all heard of this second flood as it essentially covered the entire world a mere 3 to 4 thousand years after the Atlanteans colonized Greece.]

Dual Flood

That section isn't the only revealing elements of the story. We know the worldwide flood happened AFTER the Atlantis incident for a number of reasons, but one is rooted in Plato's writing. The Atlanteans knew what was happening and began to colonize many areas including Greece and Egypt. In the second flood, there was little warning.

Plato's Flood Timing

Even Plato's timing matches the evidence and the presented timeline. I know you have been told Plato's Atlantis sank 10 thousand years ago, but that is not what his history stated. Assuming the information was obtained about 1000BC and the Egyptians were instructed for 8000 years, and Greece was instructed 1000 years before that, then the time of the sinking was between 11 and 12 thousand years ago, which is the time period we are investigating. Let's read what was said exactly.

Timeaus. 23d-24a-"Solon was astonished at what he heard and eagerly begged the priests to describe to him in detail the doings of these citizens of the past. "I will gladly do so, Solon," replied the priest, "both for your sake and your city's, but chiefly in gratitude to the Goddess to whom it has fallen to bring up and educate both your country and ours - yours first, when she took over your seed from Earth and Hephaestus, ours a thousand years later. The age of our institutions is given in our sacred records as eight thousand years, and the citizens whose laws and whose finest achievement I will now briefly describe to you therefore lived nine thousand years ago; we will go through their history in detail later on at leisure, when we can consult the records."

Greek and Egyptian Ancestors

If you noticed, the above reference also confirms that Greeks and Egyptians came from the Earth and Hephaestus. This Hephaestus character was the son of Hera and Zeus. According to legend, they didn't want to keep him, so they threw him into the sea. This could be a reference to the civilization under the sea we now call Atlantis. By that reasoning, the original Greeks may have actually descended from the Atlanteans well before any of this sinking business. On the next page we find that after years separated the two nations, there arose a huge war [similar to the wars previously addressed]. The Greek Atlanteans fought the normal Atlanteans. Possibly, the Atlanteans were fighting for their lives because right in the middle of the war, the Island sank and soldiers from both sides lost their lives in the flood.

Timeaus 25c-d-"*At a later time [after the beginning of war between Atlanteans and Athenians] there were earthquakes and floods of extraordinary violence, and in a single dreadful day and night all your fighting men were swallowed up by the earth, and the island of Atlantis was similarly swallowed up by the sea and vanished..."*

Plato's Second Book of Atlantis

While the book Timeaus was interesting, the work commonly called "Critias" gives us even more insight. Here are some excerpts. Unfortunately, Plato never finished this work.

Let me begin by observing, first of all, that nine thousand was the sum of years which had passed since the war said to have taken place between those who dwell outside the pillars of Hercules and those who dwell within. [With Plato living almost 3 thousand years ago, 10 thousand years from Plato was about 11 thousand years ago.]

Atlantis once had a greater extent than Libya and Asia and afterwards sunk by an earthquake. [The description of sinking by earthquake rather than simple flood seems to go along with other data.]

Many great deluges have taken place during the nine thousand years. [Of course one worse than the Atlantean flood occurred later and practically everyone was drowned. We will find that this terrible flood occurred about 10 thousand years ago or about a thousand years after Atlantis disappeared into the sea.]

My great grandfather, Dropidas, had the original writing, which is still in my possession, and was carefully studied by me. [This is an important part of Plato's thesis. Bringing up an ancestor as a source increases the probability of truth as one would not want to disrespect one's grandfather with a lie.]

The gods had distributed the whole Earth into portions. [This is certainly inferred by other works. As we discussed in the last book, the Nephilim, mortal angels, had initial control of the world. After a while, their half-breed offspring, Dolichocephalic headed, giant children took control. This was just before the great flood.]

Poseidon received as his lot Atlantis and begat children by mortal women and settled them in a part of the island. [Almost all ancient texts tell the same story. The super humans that the ancient Jewish writers called the Nephilimic race were considered gods. These "gods" had sexual relations and children by humans.]

In the mountains lived the Earth born men of the island. [The separation of normal humans and half-breeds is continuous throughout almost all ancient works.]

With the exception of gold, the island was esteemed with every precious metal. [If the Atlantean story was a novel rather than a true history, it is unlikely that Plato would have limited the exaltation of his utopia by limiting the most precious metal from their stores.]

If the book had been finished, we would have gained more insight into this very important trade center of the world, but he didn't finish the work and the Island did sink. The question is why did it sink? For that question it is best to revert back to more common based science and long term studies.

Flood Before the Flood

Although the last verse of "Timaeus" seems to infer it, Atlantis didn't just become submerged one day. The water level, apparently, got higher each year until it was underwater because of something we call the Wisconsin Ice Age. We can be fairly certain the Atlantic Ocean is substantially higher on average than it was 20 thousand years ago because many have been checking such things and many cross referencing methods were used to test the data. As shown below: the water has increased from between 100 and 200 feet over the past 15 thousand years and Atlantis just slipped away. The data below is not just from one study, but is a consolidation of over 24 major studies. They all say the same thing.

Isotope-oxygen data [on the volume of seabed sediments]-5 studies-*109 feet increase*

Gravitation anomalies Calculation-5 studies-*133 feet increase*

Paleo-glaciological data [on the amount of the glaciation]-7 studies-*126 feet increase*

Geomorphological data [on the ancient coastal features]-6 studies-*112 feet increase*

Isostatic effect calculations - 1 study - *167 feet increase*

The Water Came From Ice

The late Wisconsin Ice Age is timed right for the Atlantis sinking event as seen in the table below. With the beginning and end date generally exhibiting the highest water levels and the mid-Ice Age time period exhibiting the lowest water level. The golden period of the Atlantean reign would have been 18 to 13 thousand years ago. Then the water slowly rose over the cities and the island was lost as the ice melted.

Late Wisconsin Ice Age- Started 20,000 years ago—ended 11,000 years ago.

Please note how very erratic the sea height has been according to this data. The graph following shows how the water height variations may has shown up in the mid-Atlantic. The portion from about 10 thousand years ago to the present does not come from the north Atlantic wave height data, but instead is derived from events described in this book.

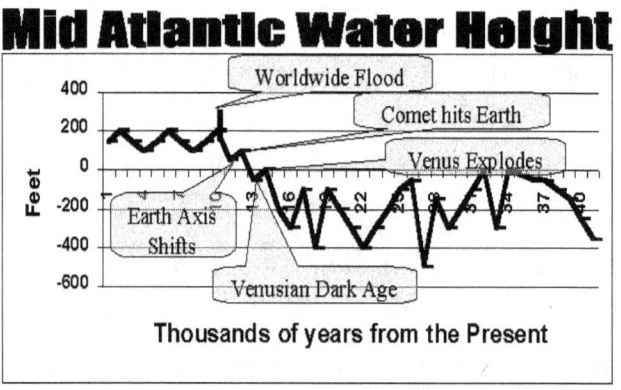

You will note from the table that about 10 thousand years ago the water level increased by 200 feet after increasing about 200 feet only a thousand years earlier. All this water level increasing happened around the time that we hear stories about a place called Atlantis and at least 6 other major civilizations that were engulfed in water and a worldwide flood that followed a few thousand years later. The following drawing shows what the world looked like with the water level 500 feet below what it now is. One thing that pops up is that the Azores becomes a huge island in the middle of the Atlantic Ocean, the Red Sea became a river, there is a huge island in the Indian Ocean, and the Mediterranean was simply a group of huge lakes connected by rivers.

The World just before the worldwide flood

Worldwide Flood
10 thousand Years Ago

Water Temperature

Possibly one can surmise that temperatures were higher if the water was deeper. Possibly Earth was pushed slightly closer to the sun, but someone decided to test the Atlantic Ocean. We can be fairly certain that the water temperature in the Atlantic Ocean abruptly increased as the Earth's axis shifted. The National Geophysical Data center provides the following information concerning the water temperature over the last 100 thousand years.

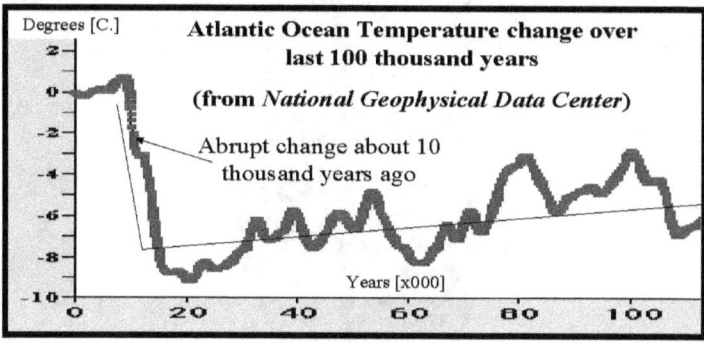

Note the change that occurred about 10 thousand years ago. The most reasonable way for that change to occur is for the axis of spin to change bringing more of the Atlantic Ocean into the equatorial region as depicted in the global views that go along with the Hawaiian Island Hot Spot data and the direction of the Carolina Bay meteorite mess.

Earth Shift Evidence

Besides giving us knowledge of a tremendous meteoritic event, the Carolina Bays provide us evidence of our last rotational shift on Earth and they even give us a good approximation of the previous axis of rotation for the Earth. That is because the density of the bays and the evidence in Australia show a "straight-line" distribution pattern that is consistent with a bombardment along the equatorial boundary. If we consider the impact density line as the "Old" equator, a shift of about 40 degrees in the rotational axis has occurred since the bombardment, as shown in the picture to below.

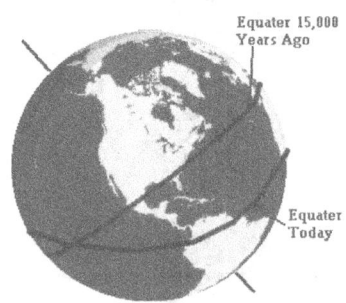

Next, the globe has been separated at this ancient equator. Note that along the equatorial path there was not much land. Also note that eastern Alaska and Siberia are well away from the Arctic Circle, which allowed huge herds of Mammoths to dine on flowers in those areas, just before the shift. The shift froze them solid.

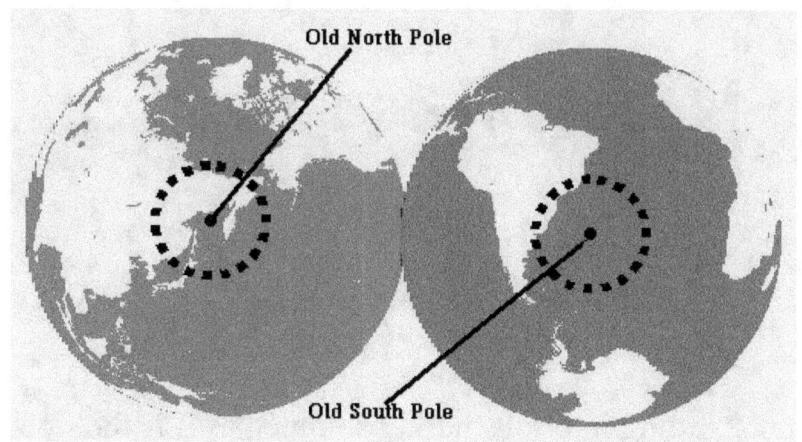

With the Atlantic Ocean more in the "Cold region" the temperature would be colder just as you would believe.

Noah's Flood Marker

I'm not getting into this marker so much as it is the flood of Noah. I think a large number of people believe in the event, but are confused with some of the anomalies like the large number of races and animal types after the flood. How could everyone fit on the boat and where did the ANAK giants reported around the world come from if Noah was the "lone survivor". Well, he was not the only survivor, but thousands and thousands of stories detail the event and scientists have silk in the Himalayas, and many other pieces of evidence that show that this event did occur and it was even more devastating that the destruction of Venus 3 thousand years before it happened. The magnetic shift data, once compressed to agree with the Ice core data easily shows both events as disrupting the axis of rotation of the earth and it marks the events about the same time as reported or described around the world. Most of the timing since this critical time can be generally believed as the Earth is more stable, the sun is fairly constant, and much of the timing is comparative since that time so I not getting into that in this book.

From the data presented, it would appear that the worldwide flood that follows was from the Earth shift. The remaining ANAK re-taught "Adamic humans" about writing, sorcery, weaponry, flying, and levitation after each of the devastating events, but even more devastation was coming and many people knew it. Pre-flood information was secured in various sites around the world and the individuals that were warned began building their rescue vessel. People knew something was wrong when the Earth shifted, but they didn't have time to react. The individual we know the most about during the final flood is Noah, but there were others. Thousands of stories about this horrible event have been told and retold for centuries. Massive mud filled graves mark the event for us, even dinosaurs were drown in great numbers during this massive flooding of the earth as some type of comet hit, added Ice, caused massive tidal-wave action that threw drowning waves over "everything". Few survived.

ANAK Survived the Flood

While many people died during the flood, the Sumerians tell us a large number escaped by staying in flying fortresses until the water subsided enough. This is why there were still ANAK and ANAKIM in the Old Testament even though they were clearly not from the lineage of Abraham. They survived the flood along with small pockets of people around the world. Some were black, some were red, some were yellow, some were brown, some were olive skinned and some were white. I know you thought the races of people magically erupted in a very short period of time after the flood along with the massive diversity of animals, but it simply did not happen. The land locked Kangaroo, for instance, did not come from the Middle East. Somehow they survived on the Island of Australia. Sorry for the rant, but sometimes people get caught up in trying to prove some esoteric religious element and forget reality. While Genesis indicates "all the people still on land were killed, it does not say Noah was the only one to live and then magically the ANAK appeared. The Bible is filled with the details of this master race of people who were worshipped as gods surviving the flood and taking control of the Land called Canaan. Here is an excerpt from the book of "Numbers".

Numbers 13:33- And there we saw the ANAK, the Anakim, which come from the giants: and we were in our own sight as grasshoppers, and so we were in their sight. [This is after the flood and after the Babel Tower incident. The time was less than 4 thousand years ago. ANAK were still on Earth and were still having giant offspring, so it is not unreasonable to believe that they lived a long, long time and somehow survived the worldwide flood.]

People in boats and flying in aircraft were able to survive. Therefore, Kangaroos sprang up in Australia and the various races of people were on the earth instantaneously after the flood10 thousand years ago, so the concept of races is both possible and obvious EVEN if there was a flood. The Sumerian texts tell us of quite a number of people who were able to find refuge in flying ships that were orbiting the earth and that they huddled in desperation hoping that the Earth would again become livable. Others were able to find floating craft to allow their survival and insure that the various human races and oddball animals would survive the time period needed to ride out the storm. By the way, Noah did not have 5 million types of animals and plants on his boat and some of the Anakim did survive or David would not have had Goliath [the ANAKIM] to kill.

The next group that somehow survived is the dinosaurs. I will have to get into this a little later, but right now let's look at the dinosaurs that survived until the flood came along and wiped them out. It was another massacre. Hundreds of thousands died in massive piles and were buried in the mud.

Dinosaurs Died in the Flood

OK! Some of the previously depicted dinosaur "creations" were made after the flood, but most of the genetically engineered monsters certainly would have lived before the flood. Ancient texts from around the world and other evidence assures us dinosaurs roamed the earth less than 50 thousand years ago. Some died in the earth shift 11 thousand years ago, but others survived until the worldwide flood of 10 thousand years ago. As the waters rose, they drowned in great numbers. When I say flood, I'm generally referring to the Noah flood, but there have been a couple other nasty ones. Anyway, here is some pretty convincing evidence.

Mass Coelophysis Dinosaurs

In the desert of the American west researchers found hundreds of complete skeletons of Coelophysis dinosaurs in a single mass grave on Ghost Ranch in northern New Mexico. All died at the same time and were quickly buried to keep predators from scattering the bones. Almost none of the 50 million buffalo that were slaughtered in the same area can be found today, but these guys all died together and were buried very quickly. The best way to bury huge quantities of massive animals is by having a massive flood.

Mass Feathered Dinosaur Grave

Scientists have discovered a mass graveyard of bird-like feathered dinosaurs related to Velociraptor in Utah, about four-and-a half feet tall, and around 13 feet long. All died at the same time and were quickly buried to keep predators from scattering the bones.

Mass Ankylosaur Grave

A species of dinosaur Ankylosaur Gastonia was discovered in a mass dinosaur grave near the base of the Ruby Ranch in Utah. All died at the same time and were quickly buried to keep predators from scattering the bones.

Mass Velociraptor Relative Grave

Dr. James Kirkland of the Utah Geological Survey discovered a mass grave with thousands of bones of a new species of dinosaur that's been deemed Falcarius Utahensis. This new species is a relative of the vicious predator Velociraptor. All died at the same time and were quickly buried to keep predators from scattering the bones.

Huge Hadrosaur Grave

One huge Dinosaur graveyard bed in Montana contains ten thousands Hadrosaurs. All died at the same time and were quickly buried to keep predators from scattering the bones.

Huge Centrosaurus Grave

In Alberta, Canada was found one of the largest mass graves of Triassic dinosaurs in the world. This massive dinosaur bone bed is 1.5-square miles in size. Scientists say that it contains thousands of Centrosaurus bones. Centrosaurus was a plant-eating, cow-sized dinosaur, with its top-of-the-head frills and rhino-like nose horn, which once lived near the Saskatchewan River. All died at the same time and were quickly buried to keep predators from scattering the bones.

Mass Plateosaurus Grave

In Zurich Switzerland, they found over a 100 Plateosaurus that died together and were buried together about one animal every 100 meters apart. All died at the same time and were quickly buried to keep predators from scattering the bones

Sinornithomimus Dongis all Buried Together

A mass grave of young birdlike Sinornithomimus dongi dinosaurs were found in China. All died at the same time and were quickly buried to keep predators from scattering the bones.

Another Mass Grave in China

The Chinese province of Shandong is home of the world's largest mass dinosaur gravesite. More than 7,600 remains have been recovered from their rocky tomb and cataloged. All died at the same time and were quickly buried to keep predators from scattering the bones.

Massive Belgium Grave

In 1878, in Bernissart, Belgium, several dozen Iguanodon skeletons were found all together in a coal mine. All died at the same time and were quickly buried to keep predators from scattering the bones. If you are thinking these dinosaurs were pretty fun to build, I think you would be right. I would imagine the Jurassic Park image presented in modern films would have been constructed as massive zoos.

Physical Evidence Flood Timing

In this book, I have used a number of timeline elements that can be pinpointed to establish a compressed or expanded timeline that fits the physical evidence. If you're thinking that my timing analysis will be weak and require manipulation of historical data you would be right, but there is plenty of evidence to strengthen this position. If we start with an initial postulate that an Earth axis shift caused the flooding the eventually flooded the entire world, we find a substantial amount of data which points to a 10 thousand year old date. This is accomplished by dating sediment deposits around the world. The evidence shows that a huge flood did occur around the world at or about that time period. Here is a small sampling of the large amount of flood evidence.

Iraq Evidence

By sediment dating of the silt layer according to presupposed dynastic timetables, gives us a date for the great flood at about 4300BC. This date is subject to assumptions that the lengths of reigns from these ancient rulers were similar to today's reigns. [It is easy to infer that the date was much, much earlier if the rulers lived longer. This will make more sense as we go along]

Mesopotamian Evidence

A marine bed was found between culture levels of ancient Mesopotamia estimated to be over 7 thousand years Old by the same criteria as mentioned above. [The probability that the period was substantially older is high given that the ruling time-periods were longer than is normally believed.]

Tibetan Evidence

A marine bed was found in the Himalayas. [This shows that the floodwaters reached high into the Himalayas, but dating the site has not been successful.]

North American Evidence

American Salt Sea and desert date to 7600 BC as the oceans covered the land. [Certainly, this could have occurred when the Atlantis episode occurred, but the wide dispersion seems to indicate the worldwide flood.]

Scotlandian Evidence

Researchers at Coventry University have determined that a giant wave flooded Scotland about 8,000 years ago. Radiocarbon dating of sediments taken from the coastline of eastern Scotland put the date of the event at about 6,000 BC.

PreEvaluation

So the evidence is again a little confusing and generally not very consistent, but the pieces all show significantly longer dates than the 5 thousand year old date commonly presented and all show the evidence suggests that there was some type of substantial flood around the world. If we add to this the more accurate dating of the earth shift and assume they occurred at roughly the same time, we have a pretty good date, but we can further investigate written documents concerning the timing of the flood to get an even better cross comparative ascertain. Before I go on, I need to bring up one last piece of more conclusive evidence that has been locked in the Ice so let's examine the more recent elements of the Ice Samples.

Ice Core Evidence

It seems like everyone and everywhere we are drilling into the icy glaciers to check on the conditions of the earth thousands of years ago. It also is apparent that the percentage of Deuterium in parts per million compared with normal hydrogen provides us with relative temperature changes in a region while the depth of the ice has been providing consistent and accurate dating. I'm not going to show all the graphs made from the dozens of studies, but it is interesting to look at two from opposite sides of the world. The first one is from Greenland while the second is from Antarctica. Low and behold, something very strange happened 9 or 10 thousand years ago. The little blimp shows some strange catastrophic event that affected weather patterns drastically for a short time. Greenland got warmer and Antarctica got colder. This may indicate that the earth axis was shifted for a few months and returned to its preshift spin which would have caused earth flooding and tidal-waves. As the dates correspond to the general flood evidence, we can surmise that the earth wobbled wildly and massive water smashed areas across the world and rain and other weather changes were significant.

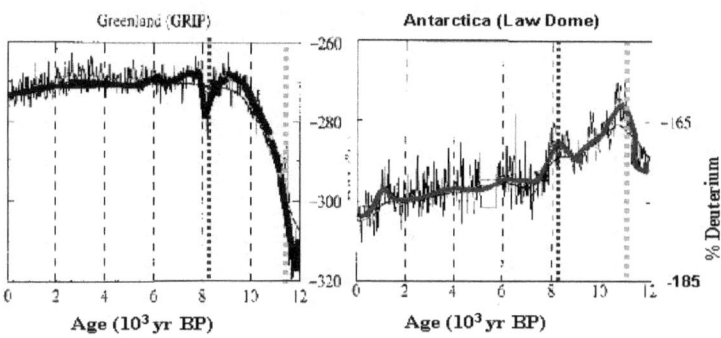

Wait a minute!! The 10000 year old event was significant, but what about the huge, long lasting event that occurred 11 thousand years ago according to the Ice evidence of the 2 sites. This must have been a time of extinction, but the other evidence does not point to this as being the flood time. We know this time as the beginning of the Holocene Epoch in scientific words. I called it the "Day Venus Exploded" in one of the books I wrote on the subject. Temperature wise, the change that occurred 11 thousand years ago shifted the thermal average and it has never returned to the old ones just like you would expect from a more permanent shift in the earth axis. We may be able to time the reigning kings to time periods by this marker just like we can the flood timing so let's look at some of the details of this disaster. This disaster affected the various kingdoms of the ancient world so it important to understand it a little.

What About the Stone Age?

Mystery of Memory Loss

In know you studied the Stone Age, The Bronze Age, Copper Age, and Modern times and this whole book sounds like mashed potatoes to all of that scientific regularity that was taught. The worst of it would be what happened after the flood 10 thousand years ago. If people were so smart, what happened??????

- Egyptians called it Zep Tepi
- Hindu called it the Age of Kali
- The PreMaya started its now famous calendar

The time was 3150BC---- Something horrible happened just before this time and I think we need to discuss it so you will see that everyone on the planet understood something massive happened.. The birth of Abram, father of the Israelites was about this time as well.

All memory of the "BEFORE TIME" was lost. The Pre Maya indicated they could no longer remember.

Hopefully you can appreciate that things seemed to be going fine for mankind after the worldwide flood disaster. One might see it as a world similar to what we are accustomed to today. One exception is that these people were even more "civilized" and with civilization comes greed for power.

War After the Flood

I'm sure they set up the same type of United Nation type of thing and soon found that it was just as useless as the one we have. They probably had a Geneva Convention type document and power hungry rulers began to ignore such rubbish. Many probably thought it was their God given right to become Imperialist and thought it would be for the GOOD of mankind that they take over larger sections of the land. We know the drill. War is almost a certainty to mankind and during this period it would have been even worse as men were smarter. This intelligence didn't guard them against destruction, but pushed them towards it. I'm sure there were many warning signs and people who campaigned for reason and empathy, but the mighty began to take over small areas at first. To do this they used technology. The book of Jasher tells us that the war was so horrible 1/3 of the population of the entire world died and another 1/3 was massively deformed.

Tower Of Babel

As you might expect, there is some controversy over when all this could have occurred. Although there is no direct date for the event or events, there are many elements, which show that a great turmoil and tragedy occurred on the Earth sometime around 3150BC. It makes sense to believe that this turmoil was the war we have been discussing. It was the war that involved the famed Tower of Babel and we will get into that aspect later. Here are some of the many indications. Any one by itself might not make us identify the time period, but after you read them all together and realize that the timing has a worldwide similarity, something will click and things will start going into place.

Indo-European Language Evidence

The noted Archeologist, Dr. Barber, indicated the following, "The first of these Eurasian expansions was proto-Indo-European. Since all the daughter languages share words for soft metals, linguists concluded that all the

Indo-Europeans already knew how to use them---therefore, they must already have entered the first age of metals, the Bronze Age, before splitting up---somewhat after 3000 BC. [The split-up was forced according to ancient texts. As people lost the capability to easily talk to one another, as we will investigate shortly, only groups in close contact with each other continued to speak a similar language and those in more remote areas changed their dialects or entire languages. Again we find it all happened around the magical 3000BC.]

Incan Long Count Evidence

The Incan calendar may also tell us when a terrible calamity occurred. The calendar is rather strange, in that it starts on 3113 BC and ends 2012 AD in groups of days known as long counts. Several researchers including the well-respected Zecharia Sitchin believe, along with me, that the 3113 BC date is none other than the climactic end of the Tower of Babel invasion. The year 2012 might also be something to consider, but that is not covered in this book.

Mediterranean Evidence

Professor Liritzis of the University of Rhodes concluded from physical evidence that a great meteor or similar event [like a nuclear blast] occurred in the Mediterranean Sea around 3150 BC.

Egyptian Calendar Evidence

The Egyptians consider the date 3150BC as the beginning of Dynasty zero or Zep Tepi. In my estimation, this zero time probably was the end of the Tower of Babel incident of this terrible war.

Genesis and Jasher Evidence

While Genesis is still in the Bible and revered as a historical truth in "2 Samuel" and "Joshua", the Biblical writers also lauded the book of "Jasher" as a great historical work as well. By using both, one can get a better picture of what the Middle Eastern world was like before and during the war.

Genesis Brief Account Chapter 11

11:1And the whole Earth was of one language, and of one speech. [Identical to many accounts around the world, before the war, people had some innate capability to understand people talking in another language.]

11:2And it came to pass, as they journeyed from the east, that they found a plain in the land of Shinar; and they dwelt there. [This Babel Tower was West of the center of Israel so it was located in Lebanon]

11:4And they said, Go to, let us build us a city and a tower, whose top may reach unto heaven; and let us make us a name, lest we be scattered abroad upon the face of the whole earth. [There was a good reason the people were worried that they would be scattered--- the world was in the midst of a terrible war.]

11:6And the LORD said, Behold, the people is one, and they have all one language; and this they begin to do: and now nothing will be restrained from them, which they have imagined to do. [This one always seems strange-no matter how many times you read it. The only thing the people had done was to "think" that making a tower would allow them to fight their enemy. Even the stupid idea that a tower could be designed to go into the heavens is absurd. Wait till you read Jasher next to find out a more complete picture of the Babel Tower and why God and some of his angels destroyed the tower.]

11:7Go to, let us go down, and there confound their language, that they may not understand one another's speech. [At first it seems that God just changed every person's language.]

11:8So the LORD scattered them abroad from thence upon the face of all the earth: and they left off to build the city.

11:9Therefore is the name of it called Babel; because the LORD did there confound the language of all the earth: and from thence did the LORD scatter them abroad upon the face of all the earth.

Jasher Explanation Chapter 7 to 10

7:46-48 And all the Earth was of one tongue and words of union, but Nimrod -was more wicked than all the men that were before him and Terah -- prince of Nimrod's host, was - very great in the sight of the king. [While Abram's [father of the Jewish nation] father, Terah, is typically depicted as a Shepard, it is unlikely that a nonmilitary man would have gained such a high place in the Nimrodian empire.]

7:51 Terah called the name of his son that was born to him Abram, because the king [Nimrod] had raised him in those days, and dignified him above all his princes that were with him. [Abram would have also been taught military tactics if he were raised by Nimrod the warring the king.]

9:19 And Abram said - these are not gods [those worshipped as gods] that made the Earth- but these are the servants of God, and Abram remained in the house of Noah – [Abram ran away from the military life. It is highly probable that the world war was already underway at this time. The gods referred to by him were the race of people that were the ruling class. They were large, very intelligent and lived long lives, but they were NOT gods. They just began to think of themselves as gods and began fighting for power and land.]

9:21 And all the princes of Nimrod said to each other, Come let us build ourselves a city and in it a strong tower, and its top reaching heaven, and we will make ourselves famed, so that we may reign upon the whole world, in order that the evil of our enemies may cease from us, that we may reign mightily over them, and that we may not become scattered over the Earth on account of their wars. [This is referencing the battles previous to this time. The wars had been so bad that they were afraid that they would be scattered all over the earth. This whole concept seems foreign to "Normal" wars where the victors would bring the captives back to a single land. Why would they even conceive that this would be done unless the whole world was at war?]

9:24 And they began - to build the city and the tower - whilst they were building they imagined in their hearts to war against God and to ascend into heaven. [Evidently, the war was beginning to focus on the land outside the Earth before the tower was built.]

9:27 They built themselves a great city and a very high and strong tower. [This thing took many years to construct. Building it in the middle of a world conflict seems bizarre unless the "building" had something to do with a weapon. From the Genesis account, we already found that there was concern that men were doing SO much. It is highly likely that this is referring to some launching site for missiles or even people.]

9:32 And God said to the seventy angels - let us descend and confuse their tongues,

9:33 And from that day following, they forgot each man his neighbor's tongue, and they could not understand to speak in one tongue, ["Forgot" is an interesting word. The brain actually could not remember how to talk to someone after

183

the "visit". It wasn't that the language he spoke was "confused" it was a loss of brain capability.]

9:35 And the Lord smote the three divisions that were there; those who said, we will ascend to heaven and serve our gods, became like apes and elephants. Those who said, We will smite the heaven with arrows, the Lord killed them, one man through the hand of his neighbor; and the third division of those who said, We will ascend to heaven and fight against him, the Lord scattered them throughout the earth. [Even after the "forgetting" stuff, God allowed or caused the death of 1/3 of the world population and another 1/3 became like primitive man that we recognize from around the world. This primitive state would have been much worse that the brain "forgetting" thing. You could say they got a second burst of whatever happened to the original people on the Tower.]

9:36 And those who were left -became scattered upon the face of the whole earth. [This is the only really surviving group of humans. Just imagine 2/3 of the population, essentially, being wiped out.]

10:1 And Peleg the son of Eber died in those days, in the forty-eighth year of the life of Abram son of Terah, and all the days of Peleg were two hundred and thirty-nine years. [This section is interesting when we look deeper into Peleg.]

More Biblical Evidence

Peleg ruled the Jewish community around 5 thousand years ago. Our interest isn't in what he did. He is really only interesting by his name. Peleg got his name from the term for "the Earth was divided" so we have to wonder what his name meant. It could mean two different things. It could represent a time when the people from the Tower of Babel were scattered around the world, but that doesn't quite fit

184

the name. A more reasonable event would be a division of 2 sets of beings on the Earth. A major division in world populations probably means war. If it means a war between the ruling race or Aryan populations against the Adamics and other followers of the Creator, then the timing is right because Peleg was the great, great, great grandson of Shem; while Nimrod, the creator of the Babel Tower, was the great grandson of Ham from a marriage outside the Adamic/ Jewish line, probably to one of the "ruling ANAKIM race". By many cross comparisons we can establish that Peleg was alive during the Babel Wars and the timing is right for him to be in power during the time when the "Earth was divided" in war—just like his name implies. If we assume that the 3300 BC date is a close date of his beginnings and the 239-year life would make the end of the war between 3071 and 3300 BC.

Magnetic Field Evidence

So we have the Bible, Inca, Egyptians, Jews, Indo-European, and Mediterranean evidence. Shouldn't there be other physical evidence as well. The answer is-- absolutely. Most of the evidence will be presented as we go through the devastation of each area of the world, but a researcher, Christopher Knight, noted a substantial perturbation in the magnetic field of the Earth that occurred around 3150BC [imagine that!]. This magnetic field shift was determined to be greater than that caused by the Comet strike of 8000BC [Noah's flood], however, both could be recognized. I don't know many things that would disrupt the earth's magnetic field, except very large weapons or another Comet strike. I'll go with the weapon thing.

Earth Thermal Cycle Evidence

Yugoslavian Astronomer, Milutin Milankovitch, determined that the earth's cyclic thermal change was altered about 5 thousand years ago. This change caused an abrupt heating to occur which was too fast of a change for his mathematical model. Afterwards, the cycle returned to a more normal cool down cycle about 4 thousand years ago. Whatever happened around 3,000BC was no laughing matter by his determination.

Tree Ring Evidence

The use of nuclear materials during this latest war is evident in the trees. Dendrochronologists have been counting tree rings for many years now and keep coming up with a terrible find. The giant sequoias and the rest of the trees we see today are not hundreds of thousands of years old, but less than 5 thousand years old. According to tree ring evidence and the article "Longevity under Adversity in Conifers," in Science, 1934, they were all wiped out at this time by some terrible catastrophe and had to begin again. It is unlikely that the worldwide flood is the catastrophe noted by the tree evidence because the flood dating points to a 10 thousand year old event. Whatever it was, it affected a large amount of the earth. My guess is a huge world war especially concentrated in North America and in India held the world in turmoil and even the plants began to die as a result of something that happened during the war.

Radiocarbon Evidence

When a team of researchers calibrated the carbon-14 dating method recently, and 25-thousand radiocarbon dates were graphed, they found something they hadn't expected. The result showed evidence of a great peak of deaths 4 to 5 thousand years ago. I don't mean just human deaths; I mean many animal types and plants as well. This either was caused by the meteors that fell to initiate something scientists like to call the Icelandic Dark Age or the World War that preceded it. Never the less, the devastation was terrible. One estimate has the destruction level during this critical time at about 20 percent of the human population. I know that is less than indicated in the Book of Jasher, but Jasher was not necessarily talking about the entire world. Now that is a war! To put it in comparison, World War I/II together caused the deaths of about 70 million or about 3 percent of the world population assuming they all died without replacements.

Jericho Evidence

I'm talking hear about the same Jericho that "Joshua fit the battle of" in the song and story. In this case, there was a horrible battle or battles that took place much earlier in time. The date of man's habitation in Jericho can be traced back to about 3200 B.C. The best illustration of this can be found in tombs. Some of the cemeteries of old Jericho have been found and the effects of the corpses there have been carbon dated to 3260 B.C. ± 110 years. One of the graves contained 113 burnt skulls arranged round the tomb's chamber. The center of the chamber contained a heap of burnt bones of the 113 bodies. There has been no correlation of burning the dead anywhere in this area. One can suppose that a huge fire such as one would expect in a war essentially destroyed the city. With that, let's look at the most famous of the Tower of Babel War cities located in the Harappan Valley between India and Pakistan.

War in India

The most noticeable devastation from the World War seemed to have happened in the area of India and Pakistan. During the war years this area was still the religious and cultural center of the world. Here is a description of some of the devastation found. Just imagine the "Giant RED skinned rouges that were against the creator God going against the normal sized, non-red skinned people.

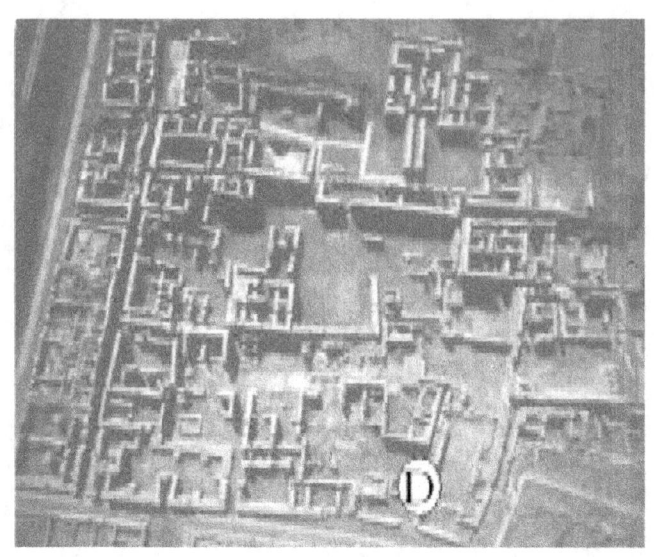

Pakistan

In the remains of Mohen-jo-Daro, shown above, or "Mound-of-the-Dead", hundreds of black lumps of melted clay pots littered the streets and skeletons were found in the street holding hands as if in complete terror during the last minutes of life. Some researchers have indicated that these skeletons were the most radioactive ever found. As you might expect, stones on walls were fused together as if a nuclear explosion occurred. Carbon dating of some of the remaining skeletons indicated that they were certainly over 2500 years old, and the site itself has been determined to be much older. The image following and the second group both show some of the many radioactive skeleton remains in the bombed city.

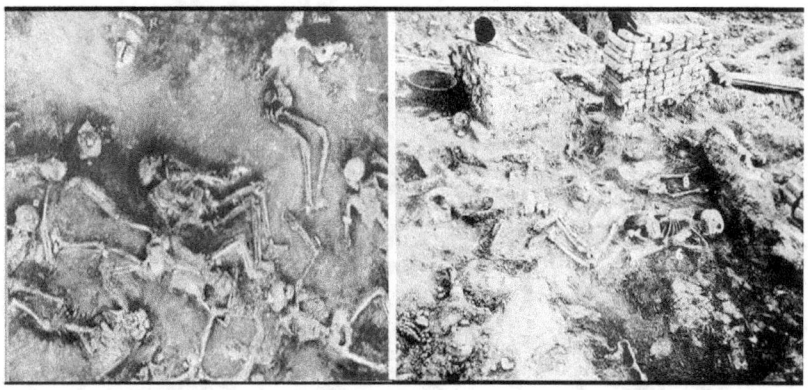

Carbon dating after a nuclear blast is hampered due to the increased carbon isotopes that are formed. It is strongly believed that the incident actually occurred over 4000 years ago during the World War. As shown below, the bodies were strewn everywhere. They had no time to flee.

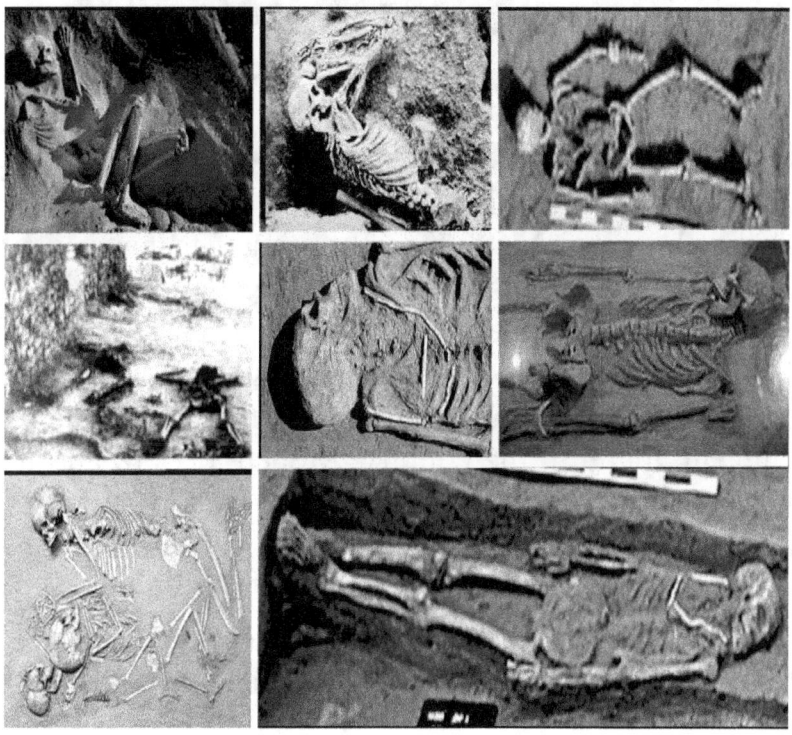

The image below is of the remains of their pottery. They had all turned to globs of glass as the intense heat melted clay and flesh during those horrible nuclear wars.

Northern India

To increase the probability of a nuclear war, an ancient city near the Rajmahal Mountains was found to have vitrified foundations and walls. These artifacts strongly suggest nuclear weaponry, intensive heat, and war. The signs in India are not the signs of some arbitrary ecological disaster, but it was what was detailed in many ancient historical documents found in the country. Mohen jo Daro and many other sites had ceased to exist as a result of assaults by manmade weapons. The only way to survive the type of destructive power presented during this time was to live underground and that is exactly what we find around the world. Unlike Hiroshima, after the apparent nuclear blast, substantial portions of Mohen jo Daro still remained.

Mohen jo Daro was only one of many cities destroyed this way just before the fateful 3150BC time remembered by all. The ruins of the valley's cities are immense. They are thought to have contained well over a million people, with a system of town planning with straight streets and rectangular blocks, as well as wide main streets like modern boulevards, and heated public baths, a network of canals, pipes and sewers, with inspection peepholes, and an efficient drainage system with a highly efficient piped water supply. Even today the ruins of a number of these cities are extremely radioactive. In Mohenjo-Daro, clays and rock were crystallized, fused or melted. Out to a distance of 60 meters or more from the general center bricks are melted on one side, indicating a blast. The excavations of the street of Mohen jo Daro revealed 44 scattered skeletons, flattened to the ground. After thousands of years, they still have a radioactivity level similar to those found in Hiroshima. Later excavation unearthed more skeletal remains in other Indus valley city ruins like Harappa, Dholavira, Lothal which number more than 300 with many vitrified or fused together masses.

Other Cities

Around the world, we find similar things. Many ancient forts in Scotland have the walls melted and the same thing has been found in France, the United States, Turkey, and many other sites as the wars were on the whole expanse of the earth.

Ape Theory

I have a theory of what happened to memories and the ability to speak to people without having to use voices as described in the Bible, many and many other ancient texts. The scientists of this time decided to help halt the wars. In their labs they built a germ or enzyme or something that was supposed to allow one winner. Unfortunately, all were losers. The affect must have almost been immediate. Soon the DNA structure of all in the world was affected. We could no longer use all of our massive brains. Some believe we could only use 10% of our brains even though that would be preposterous. Our brains would atrophy and get smaller just as has been seen with Neanderthal brain size bigger than our current brain. Some people even had children who were "Ape-like or leathery like having elephant skin as can be seen in a number of images from around the world. The following graphic shows how the Neanderthal brain size was larger than our current brain as we are slowly having smaller and smaller brain size from disuse.

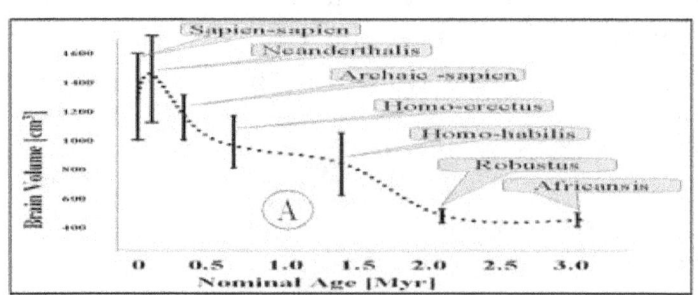

Ape People

Something turned a number of people living in society into what appears to be apes just as the Book of Jasher had indicated in his account of this war. We know this because there were many depictions of these people in ancient Cambodia, Egypt, Ecuador, Mexico and generally everywhere. I'm not going to dwell on this too long except to show you a small sampling of the disfigured people during the time just after Zep Tepi. I know it's a crazy notion, but there are plenty more images that don't have more acceptable explanations.

This first group is from Egypt showing that these unfortunate people were well regarded in Egyptian society. They mostly appear as guards.

The huge baboon guards wielding a knife were evidently; to protect the electrical things typically called Dendera tubes. While we don't know exactly what these things did for the pharaohs, we know they were extremely important. So here is my theory, for what it's worth. If you hired baboons to guard this most valuable thing, the must have been quasi-human baboons around at the time. These ape people were also depicted all over the place in hieroglyphs and statuettes showing the ape people were an important part of their society during this time.

The Aztec Ape

The Aztecs have Quetzalcoatl. Like the Egyptian ape-god Thoth was a baboon and a human. Here is a picture [left below] of the Aztec version of an ape-man god.

Panama Ape

Mayans were no different than the Aztecs. They also worshipped apes. This effigy is in the form of a footstool or platform. [Right above]

Mayans of Ecuador

In Ecuador, this monkey-man god was revered. Why would the ancient people of Ecuador revered this backward looking half human unless some of the leaders were monkey-like?

The Cambodians

In the ancient capital city of Cambodia is called Ankor-Watt. Similar baboon-men gods appear to be guarding the entrances to the temples [Next left]—Who knows?

India

In Indian history, Monkey-men soldiers fought along the side of their hero Rama, who was battling his brother who was evil. Smoke is billowing out of two of the monkeys shown, but I don't know what it means, but from the looks of their wounds, they were in trouble. The one with 12 holes is Rama.

The Lankans

In Sri Lanka, the Hindu Monkey-man god named Hanuman escaped the flames of a burning city

Elephant Theory

The Book of Jasher also indicates some of the people were turned into elephant like people. Let's look at some of the images of post Zep Tepi people that were not like the apish disfigured ones.

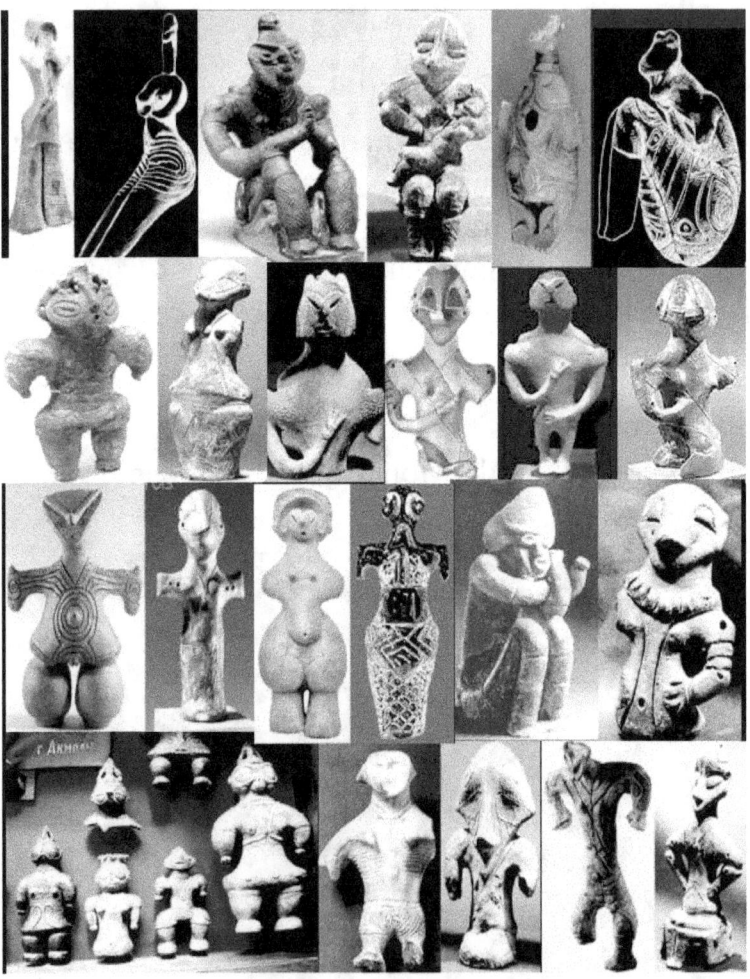

These guys had leathery skin, almost reptilian and somewhat pointed beaklike heads with large almond shaped eyes, for the most part. Disfigured people were integrated into society even when they did not look like monkeys so we can believe there were many of these people around the world. Below are a few of the places that had substantial numbers of images dedicated to this unfortunate lot of war survivors.

I brought out these images to reinforce the vision that the war was massive and the destruction was unbelievable. Something affected the very DNA structure and, most likely, we did it to ourselves. After a short time, our diminished faculties also caused a major loss of memory, skill, and capability bringing in a new Stone Age/Bronze Age/and finally another Iron Age eons after Adam's children made dinosaurs, flying machines, and weapons of war.

Even the recent timing of the war presented here has been bastardized over the years. Sometimes called the Aryan invasion, the English presented the evidence of this war as their initial onslaught into India 3 thousand years before they invaded again. Unfortunately the data they used to justify the date is now known to be describing a war between the Aryan [Iron men or Red men] and the Dravidian people of India over 5 thousand years ago. This is known because of the river system described in the documents which disappeared 5 thousand years ago.

I know this is in the opposite direction compared to what I have been presenting in this book, but it was only dated to be more recent to make the Brits feel better without ever investigation the actual time. I understand teachers are still teaching the Aryan invasion thing, but there is no truth in it. The thing they should be teaching is that 5 thousand years ago people not only had their brain size begin to shrink and they lost many of the capabilities and memories they had from ancient times, some became like animals. Additionally people began to live shorter lives.

Life Span Change

The whole concept of people losing intelligence because of some germ or outcome of a huge war is only one of the far-reaching implications of the time period associated with the Tower of Babel. There is also lifespan. Some may believe that in the ancient times people had about the same life spans as we do today, but there is so much written testimony from around the world that not having long lives is not very probable. Others believe that people used to have long life spans, but God shortened life spans to 120 years before the worldwide flood. This was never stated in any ancient documents and is an embellishment of a statement in the books of Genesis, Enoch, and Jasher that means something completely different. There are many references about humans living long periods of time up until 5 thousand years ago so we can be pretty sure there was no presto-Chango life reduction until the war we have just been reviewing.

How and Why People Live Shortened Lives

Before I go on to the evidence of humans being highly civilized and intelligent before this catastrophe, let me bring up the obvious fact that people no longer had long lifetimes and this change occurred after the tower incident.

Not Because of Generation Separation

Some have indicated that shortening of lifespan was a gradual biologic reality of our genetic code being slowly disrupted as more and more generations were removed from the initial creation of man. The problem is that our life spans seem to be getting slightly longer rather than shorter today.

Not Because of the Tree-of-Life

There are some who indicate that removal of the "tree of life" made life spans begin to reduce. The problem is that there is no evidence that man had access to this food beyond the Garden of Eden time. There would have been an immediate reduction in life span after the flood, but we find people living long lives up until the Tower. After the war, there is an immediate change to show that DNA was somehow affected.

Not Because of a Canopy

There are some who believe that a canopy of water over our planet protected us from cosmic rays and allowed us to live longer. I believe this canopy, while possible, and even probable, could not have been around as early as 40 thousand years ago and no benefit would have been given to the more recent inhabitants. If a canopy of water had somehow made people live longer before the flood and it was no longer in the sky afterwards, don't you think people would not have lived long periods of time immediately after the flood? People did live long lives for thousands of years after the flood. So we must investigate further.

Not Because of Crossbreeding

There are some who believe that cross breeding of the much longer living Nephilimic humans was the reason for the life extensions and now that those humans are gone, so are the life extensions. While that has some merit, it is my strong belief that Adam and his descendants who had long lives were not inbred. There is a huge amount of information showing their level of purity in breeding.

Not Because of the Flood

There are some who think that the historical references of the Bible, Egyptian documents, Persian documents and Babylonian documents referring to long life times halted around the time of the worldwide flood. While there is evidence of this, the probability that a flood could cause this major change in our species seems remote.

Life was Shortened During the Babel Wars

It is my belief that the actual shortening of life did not happen because of some water, increase in cosmic rays, inbreeding, or genetic decay. I believe it happened when the loss of communication was noted; when the loss of long distant viewing occurred; and when the loss of "knowledge of all things" happened and when many humans began to take on the apelike traits. That time, by most ancient sources, was the time of the Tower of Babel, some 55 hundred years ago. If there was a major war, the probability that germ warfare was used is highly likely. We use them today and it is a very effective way of taking control of areas without much bloodshed, but here's the downside.

The likelihood that something went wrong with some of the bugs also seems probable. If these bugs could mess up DNA and not let us use some of our brain as is stipulated and/or easily inferred from ancient documents and other evidence, this same "bug" could also disrupt the DNA sequencing such that life times were drastically shortened.

Whether you believe that the preflood humans had lifetimes of a mere 1 thousand years or over 6 thousand years as suggested below, it doesn't matter when arguing this point, because there was a major change 5 to 6 thousand years ago and the Tower came down about that same time. The first table is from Jewish and Biblical texts. The antediluvian ages have been expanded by a factor of 5 to keep in line with other evidence that suggests the Adam creation occurred about 40 thousand years ago. It also was modified to place the worldwide flood at about 10 thousand years ago as the physical evidence suggests. Please notice what happens after the Tower incident. On the left are the patriarchs of King Lineage and the dark blocks show the lifetimes of each.

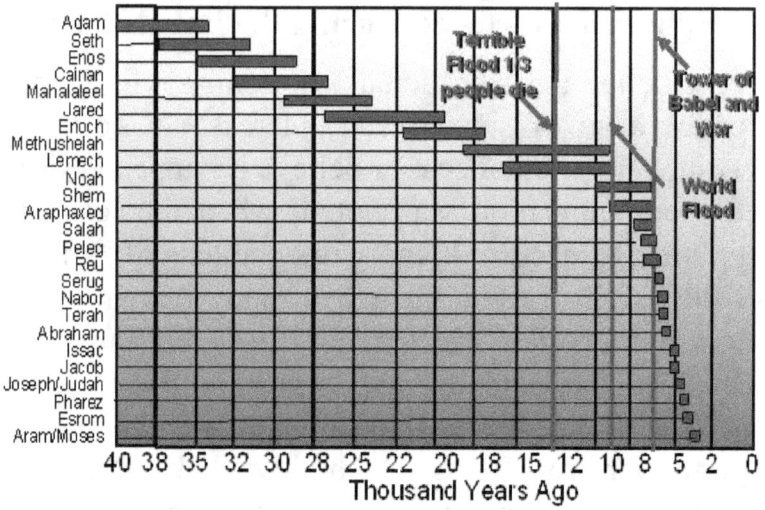

The Chaldeans and Babylonians had a similar structure of their historical timeline. In this one the King's reign time periods have be reduced to 1/12th the recorded time in keeping with the 40 thousand year common point. Again, we find a drastic reduction in ages presumably after the Tower of Babel incident. In the graph, the dark blocks are now representing the King's reign time.

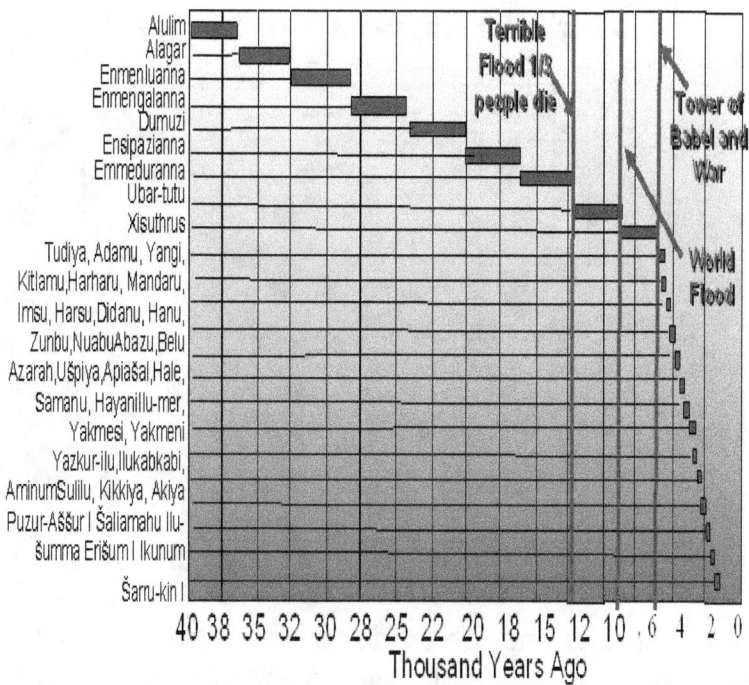

Thousand Years Ago

Like the others, the Egyptian timeline looks the same with long reigns before the flood and tower, followed by much, much shorter reigns and lifetimes after the Tower "war". In this case, the antediluvian reigns have been reduced in time by a factor of 6. Rather than showing each of the rulers after the Tower, I have clumped them by Dynasty so that each block after the Tower represents about 6 to 10 rulers. The Egyptian timeline is shown next.

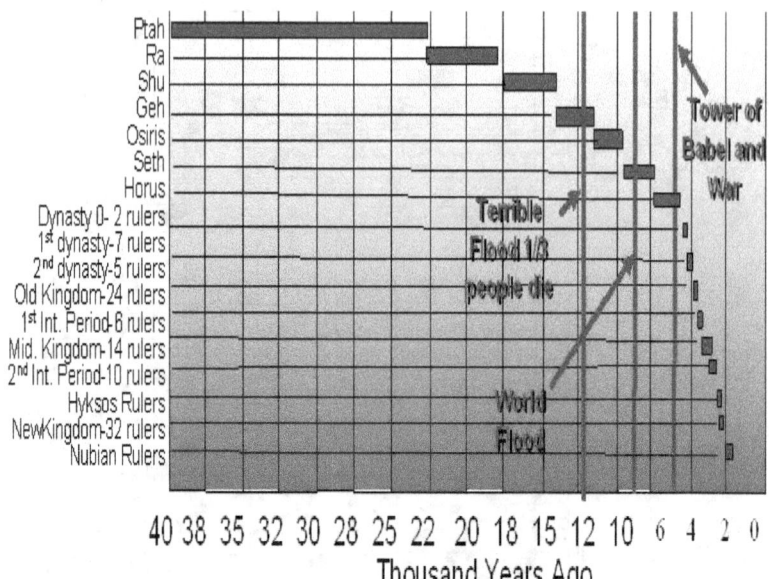

Thousand Years Ago

I believe there is enough evidence to suggest that when the Tower came down, people had shorter life times and were less intelligent. The shortened life spans are easily recognized. The intelligence limitation can also be recognized and tracked. One way to track the intelligence issue is to see how humans lived before and after the Tower incident and what I call the True Aryan War or the Babel War that immediately followed the construction of the tower. Before the war, humans had capabilities we still have not regained today. One of the capabilities lost was Genetic breeding, but the post flood people did make quite a few animals before the wars started. One of the types we have a problem with in trying to establish a reasonable time base is dinosaurs.

Dinosaurs Live Past the Flood

Behemoth

As we get into this next section we will be talking about very recent sightings of dinosaurs---well after the flood. You probably are wondering just how people could have the scientific knowledge after the flood to accomplish all of this genetic stuff as many have told you 7 thousand years ago was the Stone Age and people were throwing rocks at each other. For now, just accept what I'm saying and I promise to explain how all of this DID happen.

In the centuries that followed the Heaven Wars, we now know some of the dinosaurs and other large creatures must have survived or were regenerated. I know that doesn't sound right after you read about the extinction of dinosaurs as the earth's spin began to slow down and the dinosaurs simply got too heavy to survive, but thousands of years ago depictions of these creatures were captured. In Mexico and in Peru images of men with dinosaurs were drawn. As shown on the left below, in Peru, images of dinosaurs were often put on clay while clay models of dinosaurs and dragon-like creatures were molded in ancient Mexico, as shown on the right.

Additionally, many stories of the dragons were left by the ancient writers and artists from around the world. The images and stories indicate that, not only did the Earth inhabitants walk with dinosaurs a couple hundred thousand years ago, but also humans were in company and the company of Dragons and similar creatures up until 10 thousand years ago or less. Confrontation probably did not always have a pleasant outcome. Some of the creatures called Dragons, Leviathan, and Behemoth, were identified even in Biblical texts, but most did not survive the worldwide flood as they would not have fit on the boat and they were considered abominations. Most died during the wars before the flood and the massive destruction that happened 11 thousand years ago which we will discuss in more detail. I'm not going to go into the hundreds of stories about Dragons from all parts of the world in the history, but it should be sufficient to say that the existence or memory of those magnificent creatures was well known during ancient times. Below are some more of the pictures found in Peru which show that dinosaurs were still alive during much of the pre-flood era. The one on the left was a depiction found in a cave while the depiction on the right was a huge depiction inscribed in the Nazca plains of Peru. One thing to note about the depiction on the Nasca plain is that the drawing was done after the worldwide flood. If it was drawn before the flood, it would have completely washed away. This most devastating flood occurred about 10 thousand years ago so we can date some of the behemoths in America.

What these images and the hundreds of other similar ones show is not a mystery or an anomaly, it simply shows that genetics was mastered by people who survived the flood and recreated dinosaurs were absolutely seen as late as 3 or 4 thousand years ago.

Man almost certainly saw dinosaurs well after the dinosaurs were supposedly extinct. They were RE-created by the scientists of the day before the worldwide flood and then again after the flood according to the evidence.

It should be noted that even the Bible recounted dinosaurs surviving until about 3 thousand years ago. That means they survived the dinosaur extinction, the Heaven War and even the worldwide flood. Here are a few of the texts that describe these monsters.

II Esdras 6:48-Then he set apart two creatures the Behemoth and the Leviathan. You put them in separate places. The country of 1000 hills was given to behemoth. [A few dinosaurs survived.]

Job 40:15-24- Behold now behemoth, which I made with thee; he eateth grass as an ox. His strength is in his loins, and his force is in the navel of his belly. He moveth his tail like a cedar [There are no cedar tailed hippos, but the huge Diplodocus dinosaur had a tail worthy of mentioning]: The sinews of his stones are wrapped together. His bones are as strong pieces of brass; his bones are like bars of iron. He is the chief of the ways of God: He lieth under the shady trees, in the cover of the reed, and ferns. Behold, he drinketh up a river, and hasteth not: He taketh it with his eyes: his nose pierceth through snares.

San Antonio Behemoth

As far as more modern representations of the behemoth or land dinosaur, let's look at one that was killed in San

Antonio Texas in 1997. Supposedly, the rancher that killed the 5 foot long beast indicated that it kept on killing his chickens and even a donkey. The dinosaur had two-fingered hands and a rigid back spines along its entire back. [Next Left]

Palestrina Behemoth

This mosaic done around 100AD shows a very impressive behemoth being attacked. Maybe they were hunted like other animals a few thousand years ago. [Right]

Anasazi Behemoth

A very clear petroglyph of a dinosaur found at Natural Bridges National Monument in Utah. It is attributed to the Anasazi Indians who lived in the area from AD 400 to AD 1300. The image show a diplodocus like monster with a human nearby as shown next. This would have been well after the cretaceous extinction.

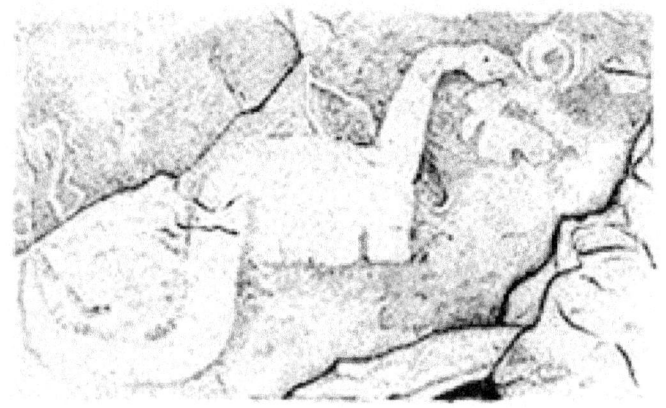

French Behemoth

Pictured below is a rock carving from Bernifal Cave in France. It shows a confrontation between a dinosaur and what appears to be a mammoth. While the carving cannot be dated to any great level, it is clearly from the post-Flood period. Clearly, these monsters survived for many years or were recreated after Noah's flood 10 thousand years ago.

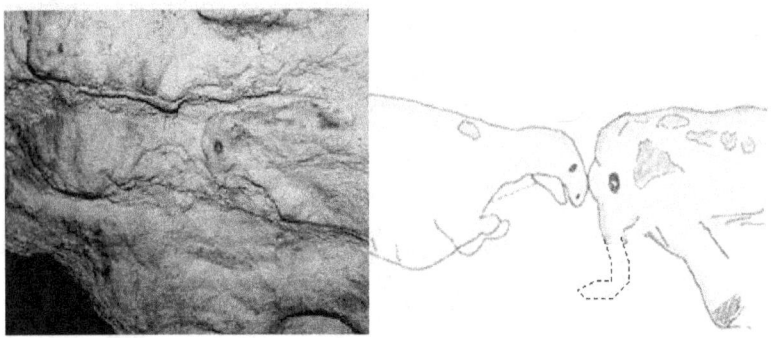

Greek Behemoth

The image following clearly shows people fighting an ancient behemoth. Most of the people of that time had never read about dinosaurs so their image of these creatures came from actual encounters. In this case, looks like the dinosaur is losing. The scientists had created or recreated these animals during their reign on earth before the flood.

Jewish Bipedal Monster

In the Jewish Synagogue Umm El-Kanatir we find art from 400 to 700 A.D. showing one of these monsters. Notice that the feet on the monster etched in stone on the following picture had three toes exactly like earlier dinosaurs. IF that doesn't look like a T-Rex with tiny arms I'll eat my hat provided that the hat is not petrified.

Western North American Monster

The Anasazi dinosaur shown on the pottery shows that these things were in the U.S.A. not too long ago. [Below Left]

Chinese Bipedal Monster

Holy Men from the Ming Dynasty [1368 – 1644 B.C.] shows one of the bipedal monsters that were seen or remembered. [Below right]

Another Chinese Bipedal Monster

As shown below, the Chinese were well aware of these monsters in the not too distant past as the image shows. The Bible called these animals either abominations or behemoth. We call them dinosaurs.

Scandinavian Two legged Behemoth

Scandinavian countries had about as many tales of dragons as anywhere in the world. One old legend describes a "reptile-like animal that had a body about the size of a large cow. Its two back legs were long and strong. But its front legs were remarkably short. And its jaws were quite large". One of the unique things about many dinosaurs was their short front legs, compared to their long, strong back legs. Many also had large jaws. Examples of dinosaurs which fit are the Edmontosaurus or Iguanodon-like. It does not seem reasonable that someone just decided that a monster should have tiny arms. They must have seen one or someone remembered stories that someone told who had seen one of these things.

The Acambaro Dinosaurs

Waldemar Julsrud, a German hardware merchant in Acambaro, Mexico, was riding his horse on the lower slope of El Toro Mountain on a sunny morning in 1944 when he spotted some partially exposed hewn stones and a ceramic object half buried in the dirt. Among the thousands of artifacts excavated were items that turned Julsrud's mansion into "the museum that scared scientists." Sculpted in various colors of clay were figurines of dinosaurs. Dr. Ivan T. Sanderson was amazed in 1955 to find that there was an accurate representation of the American dinosaur Brachiosaurus, almost totally unknown at that time to the general public.

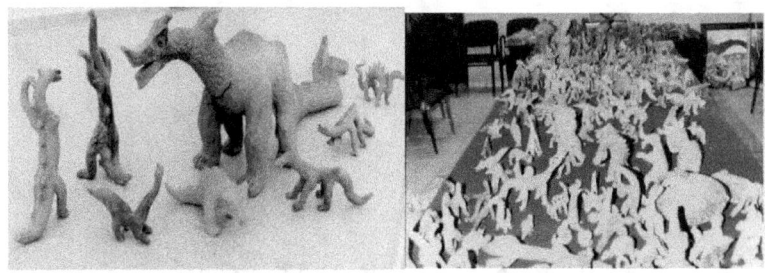

I think this show how very "normal" creating massive animals from ancient DNA was. It seems like everyone was doing it. By the way don't even think about telling me that they just found bones and decided what they probably looked like.

Mayan Monsters

There can be little doubt that the ICA stone engravings show dinosaur-like monsters that roamed South and Central America in the not too distant past. Next are just a couple of the many examples.

Colorado Monster

Huge granite carvings of the time show dinosaur-like monsters still roaming around. They were quickly sketched so as not to be eaten. Looks like the thing was a little larger than the rhinoceros depicted so we can see it was not as large as the former beasts, but there is no mistaking the dinosaur characteristics of these re-manufactured monsters.

Chinese 4 Footed Monster

If we move to the Far East, we find the same type of monster in the not too distant past. The figure following left shows one of the non-flying dragons, described in china around the first century AD.

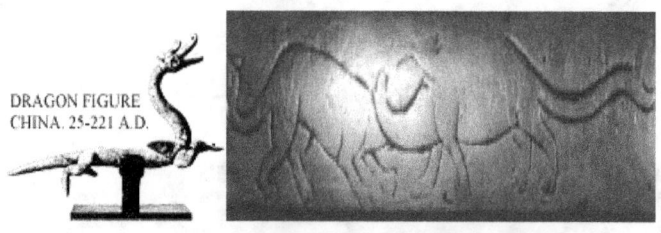

DRAGON FIGURE CHINA. 25-221 A.D.

Carlisle Cathedral Monster

Engravings in the floor of Carlisle Cathedral appear to be of dinosaurs. [See above right]They are on the tomb of Bishop Richard Bell, who died in 1496. The question is why are these brass engravings here? If they are not dinosaurs why do they have such long necks that they use to wrap around each other? The bishop's tomb is engraved with things that he enjoyed such as hunting and foliage and these dinosaur engravings...

Babylon Quadruped

One "behemoth" story from the ancient land of Sumer in Babylon tells of the hero Gilgamesh. He decided to make a name for himself by traveling to a distant land to cut great cedar trees needed for his city. He reached the forest with fifty volunteers and discovered a huge reptile-like animal which ate trees and reeds. The story simply says that Gilgamesh killed it and cut off its head for a trophy.

Irelandish Quadraped Monster

Around 900AD, an Irish writer recorded an encounter with a large beast with "iron" spikes on its tail which pointed backwards. Its head was shaped a little like a horse's. And it had thick legs with strong claws. These details match features of dinosaurs like the Kentrosaurus or Stegosaurus. They had sharp-pointed spines on their tails, thick legs, strong claws and long skulls.

French Quadruped Behemoth

The city of Nerluc in France was renamed in honor of the killing of a "dragon" there. This animal was bigger than an ox and had long, sharp, pointed horns on its head.

Peruvian Behemoth

On a rock cliff in Peru we find that these monsters were hunted by the ancient Peruvians. I redrew the animal in case you can't make it out.

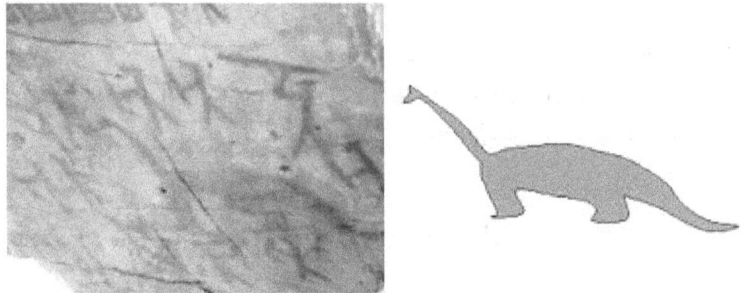

St. George Monster

Many items show how St. George killed one of these monstrous dinosaurs a number of years ago. A wonderful medieval depiction is seen at the Palau de La Generalitat in Barcelona Spain. St. George's Chapel contains an altar cloth illustrating St. George's slaying of a dinosaur.

French Chateaus Monsters

Built in the early 1500's a number of re-created dinosaurs have been depicted on wall s of Chateaus including Château Azay-le-Rideau and Château de Blois. The depictions are shown below, but they certainly look like dinosaurs before anyone had ever coined the name. There is no telling how many of these things were seen or fought against.

More are shown in the following collage.

Roman Monster

In Rome there is a similar image outside at the Church of St. Louis. This served as the national church was completed in the 1580's. There can be little mistake that a number of people saw and described these things in extreme detail.

Truth

I know some of the items I have presented are hard to believe. Some of it may still sound absurd, but let it sink in and I think you will start to see how it all starts to fit into place. One of the reasons you feel this way is that you have been taught something else all your life. You are comfortable with the version of truth that you have lived with so long. Truth is a strange word. You can have a vain truth you personally feel comfortable with and you can enlighten yourself with an absolute truth that may not make you feel good, but it will help you truly understand our world, life, existence, religion, science and historical meaning. Hopefully, this book has presented more absolute truth than vanity. Sorry for the little bit of vain truth I may have added. I believe you have a better understanding just the same. Don't get tied up because dinosaurs were around for so many years. Also don't just think that dinosaurs all died off hundreds of thousands of years ago and the rest of the anomalies can be ignored. Try to always put things together. There are no true anomalies. We simply don't know the answer. To ignore the question is a travesty.

Conclusions

There is nothing as old as you think, I would imagine. Hopefully some of this has gotten you to think. Here are a few of the things I hope, you both question and have become a little more enlightened about after reading this books.

- Nuclear decay timings is at best good for comparison of items in close approximation.
- Nuclear decay was disrupted by the wild motion of our planet, the changes in the energy from the sun, normal sunspot action, large heating sources nearby, nuclear war and a host of other things we are only now beginning to understand.
- Most of the other methods used for testing time are greatly flawed and people have been relying on those methods rather than looking at the evidence.
- By looking at the basic cyclic nature of the earth we seem to get a much more time condensed image of our history.
- The new findings of Dinosaurs being less than 40 thousand years old does not mean that the earth has only been around for that time. There is a huge amount of evidence about people who built these creatures well after they had once become extinct.
- That is not to say that people weren't on the earth during the original time of the dinosaurs. There is a mountain of evidence that tells us giant people did live during this "Golden Age".
- When Mars almost hit the earth, the nuclear timing was greatly modified the many wars fought over thousands

of years changed them and the bombardment and subsequent change in position of the Earth 11 thousand years ago modified the nuclear decay, but stabilized it to a large degree since 11 thousand years ago.

- Plate tectonics did not cause the huge mountain range that lines the "Ring of Fire" around the Pacific Ocean.

- Pangea and another massive continent I called Prestonia in this book balanced the earth before the Pacific Ocean was formed. Once scooped out, the loss of Prestonia changed a number of things on the Earth. Environmental changes became more violent, nuclear decay would have been affected, and the Earth got smaller as did just about everything on the new earth.

- We are told about a massive Heaven War. The losers became humans called the ANAK. They were worshipped as gods, but they were NOT.

- The Titans of Greek histories were not gods either, but their disappearance is somewhat of a mystery.

- Cro-Magnon and Adamic man seem to be the same group of humans.

- Neanderthal and Homo-Erectus were probably re-created by the ANAK rather than the Creator God doing it.

- Cro-Magnon was not evolved, but was created by the "Creator God' 40 thousand years ago.

- The great extinction which was previously dated at 65 million years ago only happened 100 thousand years ago.

- Archeo-magnetic, Antarctican Ice core, Greenland Ice Core, and Hawaiian Island hotspot drift timing all seem to agree and give us a better picture of our very ancient history.

- After the worldwide flood, dinosaurs were reintroduced after the worldwide flood. They have been seen all over the world in fairly recent times. The last chapter of Daniel is only about how to kill a dragon. Daniel was successful, but by his time, almost all were gone.

- There was a massive war after the flood. This was a worldwide event lasting many years. Over 1/3 of the entire planet died.
- One of the results of the war was the devolving of some people to be apelike. These people were accepted in society so there must have been many of the unfortunate people
- Zep Tepi ---3150BC became a new beginning for mankind. Around the world the same start over time was instituted.
- While Zep Tepi was a new day, we are quickly coming to a new reckoning as we are doing exactly what was done in the past to destroy mankind.

With that I hope you enjoyed the book and must end it. Sorry for the ominous end.

About the Author

Steve Preston is a long lime author of scientific, esoteric facts. His series on the creation of mankind is shown below. The series focuses on the painful truths rather than whitewashed details that make us comfortable. If you are interested in the truth instead of comfort, please continue to read and, while you are at it, review other works by Mr. Preston as shown below.

Four Part Series "Vibrational Matter"

Vibrational Matter
10-Dimentional Universe
Walk Through a Wall and Time
Meaning of Light, Life,& Death
Live and Die the Right Way [Addendum]

Eight Part Series "History of Mankind"

The First Creation of Man
The Second Creation of Man
The Creation Of Adam And Eve
The Antediluvian War Years
Man After the Flood
Life After the Babel War
A New View Of Modern History
The 20th Century To The End Of Time

Truth Series

The Truth About Dinosaurs
The Truth About The Earth
7 Destructions of the Earth
Truth About the Heaven War
Truth About Dinosaurs
Who Really Discovered the Americas?
God Didn't Make The Ape
Our Very Odd Presidents

Today's Monsters
Truth About Vampires
Living Underground

Less Ancient Works

A Closer Look At Lincoln
Adam, Lilith, and Eve
America's Civil War Lie
Ancient History of Flying
Behind the Tower of Babel
The Funny Book of Law
When Giants Ruled the Earth
Lizard People

Planet and Odd Series

When Did People Live on The Moon?
Evolution of the Planets
The Day Venus Exploded
Living on Mars
The Book Of Odd
More Oddness
Why Are There So Many Anomalies?
Stupid Science

Religious Series

Self, Soul and Spirit
Truth About the Anakim Gods
A Closer Look At Genesis
Genesis Companion
History of the World Confirmed by the Bible
Bible Inspiration Yes or No
Kingdoms Before the Flood
Four Armageddons